NATIONAL ACADEMIES *Sciences Engineering Medicine*

NATIONAL ACADEMIES PRESS
Washington, DC

# Wastewater-based Disease Surveillance for Public Health Action

AF729097

Committee on Community Wastewater-based Infectious Disease Surveillance

Water Science and Technology Board

Division on Earth and Life Studies

Board on Population Health and Public Health Practice

Health and Medicine Division

**Consensus Study Report**

**NATIONAL ACADEMIES PRESS 500 Fifth Street, NW Washington, DC 20001**

This activity was supported by contracts between the National Academy of Sciences and the U.S. Centers for Disease Control and Prevention through the National Association of County and City Officials (Contract No. 2021-100601). Any opinions, findings, conclusions, or recommendations expressed in this publication do not necessarily reflect the views of any organization or agency that provided support for the project.

International Standard Book Number-13: 978-0-309-69551-0
International Standard Book Number-10: 0-309-69551-1
Digital Object Identifier: https://doi.org/10.17226/26767
Library of Congress Control Number: 2023934388

This publication is available from the National Academies Press, 500 Fifth Street, NW, Keck 360, Washington, DC 20001; (800) 624-6242 or (202) 334-3313; http://www.nap.edu.



Printed in the United States of America.

Suggested citation: National Academies of Sciences, Engineering, and Medicine. 2023. *Wastewater-based Disease Surveillance for Public Health Action.* Washington, DC: The National Academies Press. https//doi.org/10.17226/26767.

The **National Academy of Sciences** was established in 1863 by an Act of Congress, signed by President Lincoln, as a private, nongovernmental institution to advise the nation on issues related to science and technology. Members are elected by their peers for outstanding contributions to research. Dr. Marcia McNutt is president.

The **National Academy of Engineering** was established in 1964 under the charter of the National Academy of Sciences to bring the practices of engineering to advising the nation. Members are elected by their peers for extraordinary contributions to engineering. Dr. John L. Anderson is president.

The **National Academy of Medicine** (formerly the Institute of Medicine) was established in 1970 under the charter of the National Academy of Sciences to advise the nation on medical and health issues. Members are elected by their peers for distinguished contributions to medicine and health. Dr. Victor J. Dzau is president.

The three Academies work together as the **National Academies of Sciences, Engineering, and Medicine** to provide independent, objective analysis and advice to the nation and conduct other activities to solve complex problems and inform public policy decisions. The National Academies also encourage education and research, recognize outstanding contributions to knowledge, and increase public understanding in matters of science, engineering, and medicine.

Learn more about the National Academies of Sciences, Engineering, and Medicine at **www.nationalacademies.org**.

**Consensus Study Reports** published by the National Academies of Sciences, Engineering, and Medicine document the evidence-based consensus on the study's statement of task by an authoring committee of experts. Reports typically include findings, conclusions, and recommendations based on information gathered by the committee and the committee's deliberations. Each report has been subjected to a rigorous and independent peer-review process and it represents the position of the National Academies on the statement of task.

**Proceedings** published by the National Academies of Sciences, Engineering, and Medicine chronicle the presentations and discussions at a workshop, symposium, or other event convened by the National Academies. The statements and opinions contained in proceedings are those of the participants and are not endorsed by other participants, the planning committee, or the National Academies.

**Rapid Expert Consultations** published by the National Academies of Sciences, Engineering, and Medicine are authored by subject-matter experts on narrowly focused topics that can be supported by a body of evidence. The discussions contained in rapid expert consultations are considered those of the authors and do not contain policy recommendations. Rapid expert consultations are reviewed by the institution before release.

For information about other products and activities of the National Academies, please visit www.nationalacademies.org/about/whatwedo.

## COMMITTEE ON COMMUNITY WASTEWATER-BASED INFECTIOUS DISEASE SURVEILLANCE

**GUY H. PALMER** (NAM), *Chair*, Washington State University, Spokane
**AMI S. BHATT,** Stanford University, Palo Alto, CA
**MARISA C. EISENBERG,** University of Michigan, Ann Arbor
**RAUL A. GONZALEZ,** Hampton Roads Sanitation District, Virginia Beach, VA
**CHARLES N. HAAS** (NAE), Drexel University, Philadelphia, PA
**LOREN P. HOPKINS,** Houston Health Department and Rice University, Houston, TX
**NA'TAKI OSBORNE JELKS,** Spelman College, Atlanta, GA
**CHRISTINE K. JOHNSON** (NAM), University of California, Davis
**ROB KNIGHT,** University of California, San Diego
**SANDRA L. McLELLAN,** University of Wisconsin–Milwaukee
**MICHELLE M. MELLO** (NAM), Stanford University, Palo Alto, CA
**JOHN SCOTT MESCHKE,** University of Washington, Seattle
**REKHA SINGH,** Virginia Department of Health, Charlottesville
**NEERAJ SOOD,** University of Southern California, Los Angeles
**KRISTA WIGGINTON,** University of Michigan, Ann Arbor

*Study Staff*

**STEPHANIE E. JOHNSON,** Study Director, Water Science and Technology Board
**KATALYN VOSS** (until June 2022), Associate Program Officer, Water Science and Technology Board
**ALEXIS WOJTOWICZ** (starting July 2022), Associate Program Officer, Board on Population Health and Public Health Practice
**CALLA ROSENFELD** (until May 2022), Senior Program Assistant, Water Science and Technology Board
**PADRAIGH HARDIN** (starting May 2022), Program Assistant, Water Science and Technology Board

# Reviewers

This Consensus Study Report was reviewed in draft form by individuals chosen for their diverse perspectives and technical expertise. The purpose of this independent review is to provide candid and critical comments that will assist the National Academies of Sciences, Engineering, and Medicine in making each published report as sound as possible and to ensure that it meets the institutional standards for quality, objectivity, evidence, and responsiveness to the study charge. The review comments and draft manuscript remain confidential to protect the integrity of the deliberative process.

We thank the following individuals for their review of this report:

**Kyle Bibby**, University of Notre Dame
**Thomas Burke**, Johns Hopkins University
**Michael D. Lairmore** (NAM), University of California, Davis
**Daniel C. Lang**, New York State Department of Health
**David Larsen**, Syracuse University
**Anna Mehrotra**, Water Environment Federation
**Kara Nelson**, University of California, Berkeley
**Natalie Ram**, University of Maryland
**Steven Rhode**, Massachusetts Water Resources Authority
**Helena Solo-Gabriele**, University of Miami
**Renee Street**, South African Medical Research Council
**James M. Tiedje** (NAS), Michigan State University
**Lance Waller**, Emory University

Although these reviewers provided many constructive comments and suggestions, they were not asked to endorse the conclusions and recommendations, nor did they see the final draft of the report before its release. The review of this report was overseen by **Georges C. Benjamin** (NAM), American Public Health Association, and **Joan B. Rose** (NAE), Michigan State University. Appointed by the National Academies, they were responsible for making certain that an independent examination of this report was carried out in accordance with institutional procedures and that all review comments received full consideration. Responsibility for the final content of this report rests entirely with the authoring committee and the National Academies.

# Preface

The emergence and rapid global spread of SARS-CoV-2 and the ensuing COVID-19 pandemic impacted lives and livelihoods across the world. The loss of millions of lives and the chronic sequelae of infection known as "long COVID" are the most tragic and direct disease impacts. Added to this pandemic burden are the lost years of education, the impact on families (especially those who had to take on additional childcare responsibilities), lost employment, the physical and mental exhaustion of healthcare professionals, and numerous other consequences. The world now understands "pandemic" in a real sense that no textbook could convey.

The pandemic also catalyzed innovation: rapid development, testing, and deployment of diagnostic assays, vaccines, and medications. As SARS-CoV-2 spread throughout the nation, public health agencies, universities, and municipalities began to detect and track the virus in wastewater. Although wastewater had previously been used to detect viruses and other microbial pathogens, detecting and tracking SARS-CoV-2 required developing and validating quantitative assays, triangulating wastewater levels with clinical laboratory data, and reporting results both within public health agencies and more broadly to communities. The need for emergency response led to multiple approaches from different municipalities and agencies to determine what worked best in general and for specific communities. The spirit of innovation and collective sharing of the acquired expertise reflects the "can do" character of our communities. Countless individuals donated their time and expertise to bring wastewater surveillance online as a critical tool in public health response to the pandemic.

The challenge now is to solidify this emergency response to the COVID-19 pandemic into a national system that not only continues to track the presence and spread of SARS-CoV-2 and its emergent variants but also provides near real-time data on endemic and newly emergent microbial threats for public health action. The full development and deployment of a national wastewater surveillance system can provide critical, ongoing data for public health decisions. The current report addresses the lessons learned from the COVID-19 pandemic; assesses targets and approaches for a diversity of microbial threats; and outlines a vision for a sustainable, flexible, and equitable wastewater surveillance system.

The complexity of a national system that achieves these goals requires multidisciplinary and interdisciplinary expertise. The National Academies of Sciences, Engineering, and Medicine brought together a committee with expertise in public health, epidemiology, wastewater, analytical methods, environmental engineering and microbiology, data science, and medical ethics. The committee has endeavored to examine the full range of approaches used by different municipalities and public health agencies. The responsiveness of the multiple wastewater facilities, state and local public health jurisdictions, and the U.S. Centers for Disease Control and Prevention was deeply appreciated.

The committee members brought their expertise and, importantly, their commitment to provide the evidence base for a national wastewater surveillance system. All have sacrificed their time, including evenings, weekends, and holidays, without financial compensation in this commitment. Although the ongoing pandemic impacted our ability to consistently meet in person, the committee, individually and collectively, brought their expertise, experience, and knowledge to the task. I cannot thank them enough.

On behalf of the committee, I would like to express our thanks and appreciation to the National Academies staff: Alexis Wojtowicz, associate program officer with the Board on Population Health and Public Health Practice (Health and Medicine Division); and Katalyn Voss, associate program officer, Padraigh Hardin, program assistant, and Calla Rosenfeld, senior program assistant, with the Water Science and Technology Board (Division on Earth and Life Studies). We extend a special thank you and deep appreciation to the study director Stephanie Johnson, who provided exceptional leadership throughout the study. Without her leadership and the work of the staff in planning, organizing, and editing, this report would not have been possible.

The history of public health funding in the United States, and specifically for disease surveillance, is one of emergency response to disease epidemics followed by a precipitous decline once the immediate threat has passed—only to be rebuilt with the next infectious disease event. The impacts of the COVID-19 pandemic, which touched everyone, has, hopefully, forever changed this approach. Having built on the innovation and expertise of all those who brought the wastewater surveillance system to the point where a true national system is within reach, it is a pivotal moment to ensure that it achieves its promise.

Guy Hughes Palmer, *Chair*
Committee on Community Wastewater-based Infectious Disease Surveillance

# Acronyms

| | |
|---|---|
| AFM | acute flaccid myelitis |
| CARES | Coronavirus Aid, Relief, and Economic Security Act |
| CCL | contaminant candidate list |
| CDC | U.S. Centers for Disease Control and Prevention |
| DCIPHER | Data Collation and Integration for Public Health Event Response |
| DNA | deoxyribonucleic acid |
| DUC | data use committee |
| ELC | Epidemiology and Laboratory Capacity for Prevention and Control of Emerging Infectious Diseases |
| ELISA | enzyme-linked immunosorbent assay |
| EPA | U.S. Environmental Protection Agency |
| EV-D68 | enterovirus D68 |
| FY | fiscal year |
| GLASS | Global Antimicrobial Resistance and Use Surveillance System |
| HHS | U.S. Department of Health and Human Services |
| HPAI | high pathogenicity avian influenza |
| ITS | internal transcribed spacer |

| | |
|---|---|
| NACCHO | National Association of County and City Health Officials |
| NASEM | National Academies of Sciences, Engineering, and Medicine |
| NIH | National Institutes of Health |
| NIST | National Institute of Standards and Technology |
| NSF | National Science Foundation |
| NSSIL | National Sewage Surveillance Interagency Leadership |
| NWSS | National Wastewater Surveillance System |
| PCR | polymerase chain reaction |
| PMMoV | pepper mild mottle virus |
| RNA | ribonucleic acid |
| rRNA | ribosomal ribonucleic acid |
| RSV | respiratory syncytial virus |
| RT-PCR | reverse transcription polymerase chain reaction |
| RT-qPCR | reverse transcription-quantitative polymerase chain reaction |
| SCAN | Sewer Coronavirus Alert Network |
| TRACE | Team-based Rapid Assessment of community-level Coronavirus Epidemics |
| TSE | transmissible spongiform encephalopathy |
| USDA | U.S. Department of Agriculture |
| USGS | U.S. Geological Survey |
| VADOC | Virginia Department of Corrections |
| VP1 | viral envelope protein 1 |
| WEF | Water Environment Federation |
| WHO | World Health Organization |

# Contents

# Summary

Wastewater-based infectious disease surveillance systems detect and quantify the presence of pathogen biomarkers, most commonly microbial deoxyribonucleic acid (DNA) or ribonucleic acid (RNA), that are shed by persons into a municipal sewer system. Whereas clinical laboratory testing tracks individual cases of infection, sampling and analysis at the wastewater treatment plant level (termed community-level wastewater surveillance) provides aggregate data from the homes, businesses, and other institutions that share a common sewer system. Wastewater surveillance was used effectively prior to the COVID-19 pandemic, most notably as a tool to monitor and respond to poliovirus, but the pandemic led to widespread implementation of wastewater surveillance in communities across the United States. Early in the COVID-19 pandemic, several municipalities and universities developed wastewater surveillance systems to explore the feasibility and potential public health value of SARS-CoV-2 RNA detection in wastewater. In September 2020, the U.S. Centers for Disease Control and Prevention (CDC) launched the National Wastewater Surveillance System (NWSS) in partnership with the U.S. Department of Health and Human Services to respond to the need for centralization and coordination of and communication between these efforts.

At CDC's request, the National Academies of Sciences, Engineering, and Medicine appointed a committee to review the current usefulness of community-level wastewater surveillance and its potential value for control and prevention of infectious diseases beyond the first 2 years of the COVID-19 pandemic. For the purpose of this study, community-level wastewater-based disease surveillance implies sampling at wastewater treatment plants; these treatment plants serve communities across a wide range of scales from as few as 100 to as many as 4 million people, with a median community size of 45,000 people. The committee's charge has two phases. In this report, which encompasses Phase 1, the committee was tasked to review how wastewater surveillance has been helpful to understand COVID-19 in communities and inform local public health decisions. The committee was also charged to examine the value of applications beyond the current COVID-19 pandemic, describe the characteristics of a robust national wastewater surveillance system, and discuss approaches to increase the public health impact of such a system (see Box 1-4 for the committee's complete Phase 1 statement of task). The Phase 2 study will offer a detailed assessment of technical constraints and opportunities for the NWSS, including research and technology development needs.

Wastewater surveillance has proven to be a valuable component of the nation's emergency response to the COVID-19 pandemic and will remain a critical data source for public health action in responding to COVID-19. Notably, as at-home testing increases and clinical laboratory testing and reporting decreases, wastewater surveillance for new variants and their spread will take on increasing importance. The pandemic spurred tremendous innovation and technological advances in wastewater surveillance, and ongoing knowledge development can help address gaps and improve analytical methods and data interpretation, not only for COVID-19 but also for newly emergent and re-emerging infectious diseases. Looking forward, the committee envisions a national wastewater surveillance system that is flexible, equitable, sustainable, integrated, and actionable, and recommends criteria and a process for adapting the NWSS to additional microbial threats. The committee also recommends approaches to address ethical and privacy concerns and develop a more representative wastewater surveillance system. Predictable and sustained federal funding as well as ongoing coordination and collaboration among many partners will be critical to the effectiveness of the NWSS moving forward. The committee's conclusions and recommendations are discussed below.

## WASTEWATER SURVEILLANCE FOR COVID-19

**The experience with wastewater surveillance during COVID-19 demonstrates that these data are useful for informing public health action and that wastewater surveillance is worthy of further development and continued investment.** Public health agencies that invested in collecting, analyzing, and interpreting the data found them useful to inform policy decisions (e.g., masking and social distancing rules), allocation of public health resources (e.g., testing and vaccination sites, public notification efforts), and allocation of clinical resources (e.g., staffing, hospital beds).

These data rarely stood alone but rather were frequently used in conjunction with other disease surveillance data sets (e.g., case and hospitalization counts), each with their own limitations and advantages, to decide on appropriate public health actions and resource allocations. Information on changing SARS-CoV-2 RNA levels in wastewater was shared with the public, often on dashboards, to help inform personal decision making. The launch of CDC's NWSS created an expanded network of utilities and health departments monitoring SARS-CoV-2 in wastewater, thus strengthening COVID-19 surveillance capacity and providing support and guidance for sampling, analyzing, and interpreting data for public health action.

**Wastewater surveillance data have been particularly useful for understanding SARS-CoV-2 trends and the spread of variants.** Wastewater surveillance provides a method to passively assess trends in COVID-19 burden in a community unbiased by the availability of testing or test-seeking behavior. As COVID-19 clinical testing and reported case data became less reliable in recent months due to many factors, including increased at-home testing, wastewater surveillance provided continued information on where the virus was circulating and the degree of exposure risk. Although wastewater surveillance is not currently being used as a standalone method to reliably estimate the number of community infections, SARS-CoV-2 wastewater data have correlated with case data and other conventional public health surveillance data. Depending on a number of factors, including wastewater sampling frequency, the time required for sample transport and analysis, and the time required for data reporting, wastewater SARS-CoV-2 viral trends have the potential to be reported more quickly or along a more consistent time frame than conventional disease surveillance reporting. Wastewater surveillance also provides comprehensive information on the relative proportions of known variants, and genome sequencing of wastewater samples is an effective strategy to screen for emerging variants among a large contributing population, thus providing information in advance of clinical testing data.

**The emergency response to the COVID-19 pandemic spurred innovation and rapid development and implementation of wastewater surveillance; the challenge is now to unify sampling design, analytical methods, and data interpretation to create a truly representative national system while maintaining continued innovation.** Early challenges in initial surveillance sites focused on analytical capacity, sampling design, and data interpretation. The pandemic inspired innovation, which led to a diversity of approaches and methods rather than a single standard method. To date, sites within the NWSS have been based primarily on wastewater utility and public health jurisdiction willingness to participate, including volunteering time and resources, and thus do not comprise a representative national system. Importantly, participating sites have successfully built new partnerships across professional communities with limited prior interactions, spurring innovation and increased efficiency. The challenge is to formalize these roles and partnerships and ensure national representation with best practices for collection, analysis, and use of the data.

## VISION FOR A NATIONAL WASTEWATER SURVEILLANCE SYSTEM

**Wastewater surveillance is and will continue to be a valuable component of the nation's strategy to manage infectious disease outbreaks, including continued surveillance of SARS-CoV-2 variants, resurgences of known pathogens, and newly emergent pathogens.** The emergency establishment of wastewater surveillance has proven its value, and the efforts at local and national scales to establish the NWSS provide a solid basis for expanded applications. Infectious diseases, whether endemic, seasonal, newly emergent, or re-emergent, are dynamic and never fully predictable. The high likelihood that SARS-CoV-2 variants will continue to emerge and circulate is alone a strong rationale to maintain and strengthen a national wastewater surveillance system. The recent use of wastewater surveillance for poliovirus and mpox in mid-2022 illustrates the advantages of a maintained national system for detecting re-emerging pathogens and pathogens recently introduced into the United States.

**To achieve its goals, a national wastewater surveillance system should be flexible, equitable, integrated, actionable, and sustainable.** Flexibility includes the ability to track multiple pathogens simultaneously and pivot quickly to new threats. A national wastewater surveillance system should be as equitable as possible across population demographics, with efforts to engage underrepresented communities and extrapolate findings, where feasible, to unsewered communities. Integration, including coordination and collaboration across multiple partners (e.g., utilities, laboratories, and public health agencies) and triangulation of data from different disease surveillance systems, ensures effective data interpretation in support of public health decision making. For the information to be actionable and inform decisions about clinical and public health resource allocations as well as policy decisions, it must also be timely, available, reliable, representative, and interpretable. Finally, the system needs to be fiscally and operationally sustainable. Although the NWSS supports both local and national public health decision making, a sustainable national wastewater surveillance program may not serve every locality's objectives but should allow for locally funded initiatives, such as pilot surveillance of a pathogen of emerging regional concern.

**When evaluating potential targets for future wastewater surveillance, CDC should consider three criteria: (1) public health significance of the threat, (2) analytical feasibility for wastewater surveillance, and (3) use-**

**fulness of community-level wastewater surveillance data to inform public health action.** Applying these criteria to known and emergent/re-emergent pathogens of concern can guide strategic allocation of effort and resources. Assessment of the public health significance of a microbial threat is important to develop and maintain a system that is responsive to current public health needs. Assessment of the feasibility to detect a specific pathogen in wastewater for disease surveillance is necessary to determine technical readiness and can also drive research or technology development for microbial threats that meet the other criteria. Finally, it is critical that the value of wastewater surveillance information for a given pathogen be considered in the context of the broader universe of surveillance approaches so as to maximize the use of resources to inform public health action (e.g., allocation of clinical or public health resources). Candidate pathogens will need to be re-evaluated periodically as scientific knowledge, technology, and infectious disease risks evolve.

**Temporal and spatial resolution of the NWSS sampling program should be subject to intentional design, informed by rigorous and iterative analysis of data for prioritized pathogens.** Collaborative and frequent analysis of incoming NWSS data is essential to determine the spatial and temporal scales of sampling and analysis needed, both for effective COVID-19 monitoring as well as detection of emerging pathogens. Temporal and spatial resolution should be regularly re-evaluated to ensure the system is capable of detecting meaningful change with sufficient lead time needed to inform public health action. CDC should also give careful attention to the need for more representative sampling for prioritized use cases. Currently, the system consists of localities, tribes, and states that were willing and able to participate during a pandemic emergency, and this current distribution of sampling sites might not be representative of the range of demographic and geographic characteristics desired in a national network, nor equitable, optimally actionable, or sustainable. Because 16 percent of the U.S. population resides in unsewered communities, wastewater surveillance in and of itself cannot be fully representative of the population but should be viewed as one key component of a national infectious disease surveillance system.

**CDC should take additional steps to bring the benefits of wastewater surveillance to critical areas not addressed by the NWSS.** The committee identified three steps that CDC could take to ensure that resources expended on wastewater surveillance systems are not distributed inequitably. First, CDC should create a comprehensive outreach program to provide information to selected public health officials and utility personnel in localities that are not currently using wastewater surveillance about the potential benefits of joining the national system. Second, CDC should reduce financial and staff capacity barriers to joining the system. CDC could reduce barriers by providing continued and expanded funding to state, tribal, local, and territorial health departments and utilities and by creating an easily operable data management and analysis system wherein local wastewater surveillance programs can easily transmit their samples and data for centralized analysis and data visualization (see Chapter 4). Finally, because some areas that are important to understanding national infectious disease transmission will remain outside the wastewater surveillance system even with these resources in place (e.g., in unsewered areas), CDC should assess whether tools can be used to extrapolate data from monitored regions to estimate disease burden in areas without wastewater surveillance. CDC and local health departments should also maintain robust infectious disease surveillance programs using other sources of data on disease trends and provide public education about how to interpret wastewater data alongside other indicators.

**As part of a national wastewater surveillance system, strategic incorporation of sentinel sites is recommended as a mechanism for early detection.** Sentinel sites should be intentionally selected to monitor for specific emerging pathogens at their points of entry into human communities. Sites that can directly inform community wastewater-based surveillance, especially as related to emerging pathogens, will provide important and distinct benefits in the context of a national surveillance network. Such sentinel sites could include wastewater surveillance at major international airports with a large number of global travelers to detect emerging pathogens and antimicrobial resistance genes. Sentinel monitoring at ports of entry could allow early detection of emerging pathogens entering the country that otherwise may be too dilute to detect at the community scale. Wastewater treatment plants with zoos or major livestock farms that contribute to its sewer system could also serve as valuable sentinel sites to detect the emergence and transmission of zoonotic pathogens. Developing useful sentinel sites will require careful planning and thoughtful experimentation with site selection, program design, and data interpretation based on the pathogen(s) of interest. Sentinel sites are a cornerstone of any public health system, and the NWSS should seek to incorporate these sites in a way that will ensure the surveillance system is nimble and adaptive as needed to address emerging threats.

## STRATEGIES FOR ACHIEVING THE VISION

**CDC should develop an open and transparent process for prioritizing targets for wastewater surveillance.** Selecting future targets for wastewater surveillance is a challenging endeavor that balances potential health benefits against resource investments and the capabilities of existing technology. CDC would benefit from an independent external advisory panel, with representation from industry, academia, and public health, to provide periodic guidance and input to this process and ensure that the latest advancements in science and technology are considered. The external advisory panel could also provide rapid consultation in future pandemic emergencies. Public input to the process is important because the community should have the opportunity to have concerns heard and considered before a final plan is implemented.

**Although the committee judges that the benefits of responsibly managed wastewater surveillance outweigh the associated ethical concerns, CDC should address privacy concerns through clear public communication and by convening an ethics advisory committee.** CDC should develop and disseminate additional public communications designed to inform the public about the data generated in wastewater surveillance and how these data are used. In addition, CDC should empanel a standing ethics advisory committee to recommend guidelines about the conditions under which wastewater data may be shared with others and to evaluate future expansions of data collection and data access. It is desirable for academic and industry partners to be able to conduct and contribute analyses of wastewater data, which requires responsible data sharing. The ethics committee, which could be modeled after existing data use committees, should create a formal process for executing data use agreements to help address privacy concerns and alleviate burdens in managing data sharing at a local level. Furthermore, if the prospects for identifying individuals in wastewater data strengthen over time, or if any agency or private-sector organization expresses interest in using wastewater data for purposes other than infectious disease surveillance, this body should re-evaluate the balance of health benefits versus risks associated with data sharing and any proposed expansions in data collection and data linkage. There should be a strong firewall maintained that precludes use of data by law enforcement. In performing its work, the ethics body should consider whether steps are needed to help avoid stigmatization of particular communities or to build further buy-in to wastewater surveillance among members of particular communities.

**The effectiveness of the NWSS will depend on predictable and sustained federal investments.** The COVID-19 pandemic emergency spurred many researchers and utilities to volunteer their labor and donate resources in support of the effort, but the vision of a sustained national wastewater surveillance system necessitates a shift from volunteerism to a strategic national plan with well-defined roles supported by federal investments. Federal funding is needed to continue to advance sampling and analysis methods and data analysis tools to improve data quality, comparability, and actionability. Predictable funding is also essential to maintain the workforce capacity and institutional knowledge to sustain a well-functioning wastewater surveillance system that is useful to public health agencies and to support an effective system for data management and interpretation for all public health agencies.

**Close coordination among public health agencies, analytical laboratories, and wastewater utilities is essential to generate reliable data and support appropriate data interpretation and use.** CDC's Communities of Practice for wastewater utilities, laboratory personnel, and public health practitioners provide valuable support for coordination within each of these fields, and CDC can work with these communities to establish expectations for coordination and collaboration with other agencies. State, tribal, local, and territorial public health agencies should also work to strengthen relationships across these partners—for example, by encouraging biweekly meetings with staff from the public health agency, the analytical laboratory department, and the wastewater utility, as appropriate, in support of data interpretation. CDC, as the nation's health protection agency, should continue to lead the coordination of the many federal partners in support of this effort.

**Because the function of the NWSS depends on the participation of wastewater utilities, CDC and local public health agencies should continue to strengthen relationships with wastewater partners.** CDC should continue to work to improve the connections between wastewater utilities and local, state, tribal, territorial, and federal public health agencies, beyond what is currently provided in the Communities of Practice. At a federal level, CDC could set expectations and standards of practice that utilities be engaged as full partners, with compensation for their participation and education and data sharing to ensure that the utilities see the value of their contributions. Local public health agencies should work to build relationships with utilities, who can also provide important expertise essential to developing sound sampling designs and accurate data interpretation.

**Looking forward, academia and the broader scientific community are essential to drive innovation in sampling, laboratory analysis, data management and interpretation, and public communication.** CDC is commended for launching two initial Centers of Excellence, which will help support targeted research and training. In addition to the Centers of Excellence, CDC should engage the scientific community around specific sampling, analytical, and data management needs through funding mechanisms such as the CDC Broad Agency Announcements. Academic and other research laboratories could provide needed training, and an NWSS workforce needs study would help ensure that a trained workforce can meet current and future needs.

# 1

# Introduction

Rapid detection of recognized or emerging infectious disease outbreaks is essential for timely public health response. The COVID-19 pandemic illuminated the strengths and limitations of the U.S. public health infrastructure, particularly the challenges to implementing widespread clinical testing, tracking asymptomatic infections, and anticipating community disease outbreaks. During the COVID-19 pandemic, wastewater surveillance[1] gained traction as an additional epidemiological tool to monitor trends and anticipate disease incidence in communities.

Wastewater surveillance systems collect samples of untreated municipal wastewater that are then analyzed for the presence of biomarkers of infection, most commonly pathogen deoxyribonucleic acid (DNA) or ribonucleic acid (RNA) that are shed by infected persons (see Box 1-1; Figure 1-1). Whereas clinical laboratory testing and health services track individual cases of infection, testing for a pathogen at a wastewater treatment plant (also known as community-level wastewater surveillance) provides aggregate data from an entire community sewershed (i.e., the community population consisting of homes, businesses, and other institutions that share a common sewer system or drainage area). It does not track or identify

[1] In this report and more broadly across the field of public health, "wastewater surveillance" describes the ongoing collection, analysis, and interpretation of and response to data related to the transmission of pathogens in wastewater for public health purposes. The committee acknowledges that the word "surveillance" is a charged term also used in other contexts to describe careful watching by the police, although that use is not intended in this report. See Chapter 4 for a discussion about privacy considerations associated with implementing a national wastewater surveillance system.

**BOX 1-1**
**What Is Actually Being Detected in Wastewater Surveillance?**

Theoretically, the actual pathogen—be it a virus, bacteria, parasite, or fungus—can be directly detected in wastewater, either by direct cultivation or following an enrichment step. In practice, this approach is limited by the low concentration and low viability of most pathogens in wastewater and the expense of direct isolation. Consequently, current pathogen detection is based on biomarkers—targets that are specific to a pathogen and amenable to detection independent of actual pathogen viability.

The most commonly used biomarker is pathogen-specific nucleic acid, either RNA or DNA. Polymerase chain reaction (PCR) methods can be used to amplify targeted genetic fragments from known organisms, mitigating, at least in part, the challenge of low pathogen concentration in wastewater. Amplification can be highly specific and quantifiable and uses instrumentation and technical expertise that are widely available. Importantly, the amplification step can be modified as needed for new pathogens or variants, assuming targeted assays are available, and the end product can be sequenced for confirmation of pathogen identity or identification of genetic changes associated with a new variant. Genes encoding resistance to antibiotics can be detected in bacteria using the same amplification techniques. The distinction here is that key determinants of antibiotic effectiveness are being detected rather than the pathogen itself. Metagenomic sequencing methods enable the identification of previously unknown variants of known pathogens (Karthikeyan et al., 2022; see Chapter 2). These approaches do not require that the pathogen be viable.

Additional approaches can directly detect other pathogen components, such as proteins and lipids, which can be used as biomarkers. Although methods based on these alternative biomarkers are less commonly used due to their relative inability to detect low concentrations, they may be required for certain infectious diseases (e.g., prion diseases) and may benefit from ongoing research to improve detection sensitivity.

For uniformity in this report, the term "pathogen detection" (e.g., detection of SARS-CoV-2) is used with the understanding that the actual detection is a pathogen-specific biomarker (e.g., an amplified segment of SARS-CoV-2) or, more broadly, biomarkers of microbial threats such as bacteria carrying genes encoding resistance to antibiotics.

infectious disease for an individual person or household; rather, it detects the presence and changing quantities of a pathogen within the larger community. In the United States, 84 percent of households are connected to a wastewater treatment plant (U.S. Census, 2022). The sizes of communities served by an individual wastewater treatment plant can range widely, from very small plants that serve as few as 100 people to large plants that serve a few million people, with a median of approximately 45,000 people (A. Kirby, CDC, personal communication, 2022). The remaining unsewered

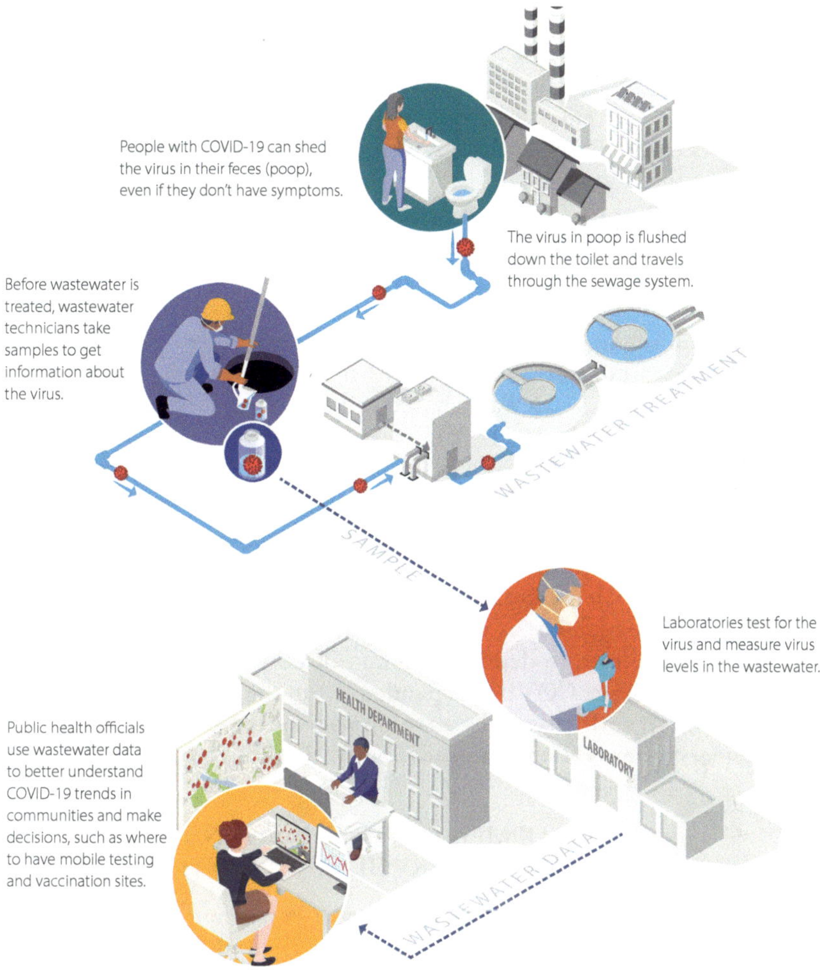


**FIGURE 1-1** Components of a community-level wastewater surveillance system. Infected persons can shed biomarkers of infection (see Box 1-1) into the wastewater system through feces, urine, saliva, and other sources. Household wastewater is discharged into the sewer system and collected at the inflow to the wastewater treatment plant, where sampling occurs. The sample is then transported to a laboratory where it is analyzed, and the data are analyzed and published on internal- or external-facing dashboards. These data are then used by state, tribal, local, territorial, and national officials to support decision making on public health interventions, and the distribution of resources and support public communication.
SOURCE: Adapted from https://www.cdc.gov/healthywater/surveillance/pdf/Wastewater-COVID-infographic-h.pdf.

population is not directly addressed by this epidemiological approach, although some members of this population regularly commute to sewered areas for work, school, or other activities.

Wastewater surveillance detects the genetic biomarkers of disease agents that have been discharged into a sewer. The measurement is inherently an indicator of the magnitude of the agent's loading to wastewater, which can be interpreted to understand the prevalence of infection in a community. Wastewater surveillance can capture pre-symptomatic cases as well as infections across the spectrum of disease severity, including asymptomatic cases that may not result in an infected individual seeking medical care. However, interpretation of data may be confounded by variability in shedding levels and patterns, pathogen stability, and other factors.

The concept of wastewater-based epidemiology first emerged in the 1940s (Paul and Trask, 1941; Paul et al., 1940). A variety of wastewater surveillance initiatives occurred in the 1990s, 2000s, and 2010s, including the use of surveillance in global polio eradication efforts (see Box 1-2), to detect prevalence of the flu, and monitor the use of pharmaceutical and illicit drugs (Safford et al., 2022). The experience of applying wastewater surveillance to poliovirus and coronavirus has highlighted the unique value of wastewater surveillance as well as potential limitations to its application. (See Chapter 2 for an in-depth discussion of how wastewater surveillance has been useful in the COVID-19 pandemic.) In the remainder of this chapter, the committee discusses the development of the National Wastewater Surveillance System (NWSS) and the motivation for this study.

## DEVELOPMENT OF THE NATIONAL WASTEWATER SURVEILLANCE SYSTEM

During the COVID-19 pandemic, broad interest in the potential usefulness of wastewater surveillance emerged. This led to the independent development of many local (e.g., sewershed and sub-sewershed; see Box 1-3) and hyperlocal (e.g., building or institution-based) wastewater surveillance efforts. As proof-of-concept was established for the feasibility and potential public health value of SARS-CoV-2 RNA detection and variant sequence identification, the implementation of these systems expanded. Wastewater surveillance was deployed at several locations across the United States and internationally to forecast and monitor disease outbreaks (see Table 1-1) and was found to be effective in capturing information about both asymptomatic and symptomatic infections as well as in predicting outbreaks (see Chapter 2 for more detail). Recognizing the need for centralization and coordination of these efforts, the U.S. Centers for Disease Control and Prevention (CDC) launched the NWSS in partnership with the U.S. Department of Health and Human Services (HHS) in September 2020. The NWSS

**BOX 1-2**
**History of Wastewater Surveillance for Poliovirus**

Given the global vaccination campaign to eradicate polio, a single instance of poliovirus detection triggers the need for public health intervention. Wastewater surveillance has been a critical tool in tracking poliovirus because, in its absence, asymptomatic poliovirus shedding can remain undetected and allow community spread until detected through clinical cases of acute flaccid paralysis (which occurs in ≤0.5 percent of infections).[a] Detection through wastewater triggers targeted screening of the community, which is more efficient and cost-effective than continuous large-scale population-based screening of individuals.

Isolation of poliovirus from urban sewage was first reported in 1940 (Paul et al., 1940). Paul and Trask (1941) completed the first study using wastewater surveillance for poliovirus, and building on this study, Melnick (1947) used wastewater surveillance to understand transmission of polio in a population. Even though clinical surveillance of acute flaccid paralysis was considered the gold standard for polio surveillance, throughout the latter half of the 20th century, numerous studies reported the use of wastewater surveillance in addressing poliovirus outbreaks (Adu et al., 1998; Böttiger and Herrström, 1992; Gershy-Damet et al., 1987; Horstmann et al., 1973; Manor et al., 1999a,b; Marques et al., 1993; Pöyry et al., 1988; Thraenhart et al., 1977; van der Avoort et al., 1995; Zdrazílek et al., 1971). By the early 1990s, wastewater surveillance was used to effectively monitor and respond to polio outbreaks with targeted public health interventions and a robust vaccine campaign in Israel and the Palestinian Authority (Hovi et al., 2001; Manor et al., 1999b; Ranta et al., 2001).

Since the early 2000s, systematic environmental surveillance for poliovirus has been performed in many countries that were endemic for polio or had risk of re-importation. The global polio eradication initiative has supported wastewater surveillance as a tool since 2013, starting in 5 countries and growing to more than 550 sites in 45 countries for routine wastewater surveillance (WHO, 2022). Wastewater surveillance continues to be an effective tool not only for fighting endemic poliovirus and re-importations but also for addressing vaccine-derived poliovirus outbreaks. The introduction and local spread of a type 2 vaccine-derived poliovirus in London was identified by wastewater surveillance (Klapsa et al., 2022; Wise, 2022). This virus was subsequently genetically linked to a strain implicated in a Jerusalem outbreak (Zuckerman et al., 2022) and a case of acute flaccid paralysis and wastewater detection in New York (Link-Gelles et al., 2022; Ryerson et al., 2022).

---

[a] See https://www.cdc.gov/polio/what-is-polio/index.htm.

## BOX 1-3
## Sewersheds and Sub-sewersheds

A sewer network can have complex configurations depending upon the individual location. The geographic area serviced by a network of pipes (sewers) feeding into an individual wastewater treatment plant is termed its "sewershed." Sewersheds can range in size from very small to very large. Wastewater treatment plants that submit samples to the National Wastewater Surveillance System serve populations that range in size from 100 to 4 million people, with a median of 45,000 people (A. Kirby, CDC, personal communication, 2022).

The scale of a wastewater treatment plant and its sewershed are influenced by a number of factors including population size and density, geopolitical boundaries, topography, and technology. Some large urban areas are served by one or more large wastewater treatment plants (e.g., the Metropolitan Water Reclamation District of Greater Chicago operates three treatment plants that each collect and process wastewater from more than 1 million people, along with four medium-sized plants [CDPH et al., 2022]). Other urban areas are served by numerous smaller treatment plants; for example, Houston Water operates 39 wastewater treatment plants for its 2.2 million customers.[a] Thus, as demonstrated in these examples, wastewater surveillance sampling at treatment plant inflow can provide quite different levels of spatial detail.

Within a sewershed there may be subareas, each serviced by a network of sewers, termed a "sub-sewershed." An example is shown schematically in Figure 1-3-1 for Jefferson County, Kentucky (Holm et al., 2022b), which has five sewersheds (i.e., five individual wastewater treatment plants) and several additional sub-sewersheds illustrated by the individual shaded areas. Many sub-sewersheds can be sampled, as in this case, by sampling at individual manholes to large sewers (shown as circles) or at the location of pump stations (shown as squares). Pump stations are the locations of convergence of sewers in a sub-sewershed in order to pump the flow to greater elevation due to the topography of the service area. Spatial resolution in wastewater surveillance is discussed in more detail in Chapter 2.

---

[a] See https://www.houstonpublicworks.org/houston-water.

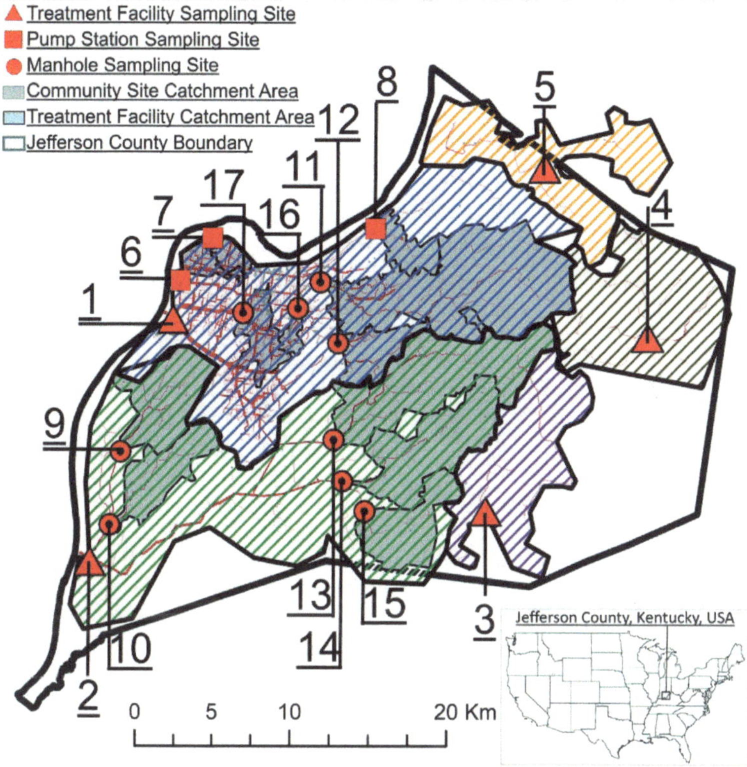


**FIGURE 1-3-1** Map of sewersheds and sub-sewersheds used for SARS-CoV-2 in Jefferson County, Kentucky.

NOTE: Triangles are wastewater treatment plants. Shaded areas represent sewersheds (delineated by heavy black borders) and sub-sewersheds. Circles are individual sub-sewershed sampling locations.

SOURCE: Adapted from Holm et al. (2022b) with permission from the Royal Society of Chemistry.

**TABLE 1-1** Selected Examples of Ongoing Wastewater Surveillance Programs as of July 2022, Including Both Community and Institutional Scales

| Wastewater Surveillance Program | | Unique Aspects of the Program |
|---|---|---|
| **State Programs** | Missouri | Missouri was one of the first states to initiate wastewater surveillance testing. This project is a collaborative effort among the Missouri Department of Health and Senior Services, the Missouri Department of Natural Resources, and the University of Missouri. This project tracks SARS-CoV-2 viral load in the wastewater of more than 100 participating community water systems across Missouri.[a] |
| | Ohio | The program involves large-scale collaboration among multiple state agencies, the U.S. Environmental Protection Agency Office of Research and Development, and numerous universities. Wastewater treatment plants are sampled throughout the state. The state developed a public-facing dashboard that depicts trends in wastewater results. |
| **Local Programs** | Houston, Texas | The city collects wastewater samples from the 39 wastewater treatment plants within the city, as well as at lift stations within the sewershed and individual facilities. The city uses the wastewater data, along with other data sources such as individual clinical testing results and vaccination rates, to identify ZIP code-level "hot spots" for targeted public health intervention. The data are also used to monitor for early alerts of waves alongside other data such as emergency visits, general hospital bed use, and intensive care unit bed use. |
| | Tempe, Arizona | The city quickly developed and implemented a wastewater surveillance program for SARS-CoV-2 by building off its existing opioid wastewater surveillance program. The City of Tempe, in partnership with Arizona State University's Biodesign Institute, generates and uses sewage data to inform city decisions and operational strategies. |
| **Privately Funded Programs** | San Francisco Bay Area, California | The privately funded Sewer Coronavirus Alert Network project is led out of Stanford University in collaboration with an industry partner. The project analyzes daily primary settled solids from 11 wastewater treatment plants in the San Francisco Bay Area, serving approximately 10,000 to more than 1,000,000 people. Originally focused on SARS-CoV-2, the project now reports the wastewater levels of a number of pathogens.[b] |
| | New Haven, Connecticut | The Rothberg Fund supports a SARS-CoV-2 wastewater surveillance program at New Haven, Connecticut's wastewater treatment facility, with joint efforts by Yale University and the New Haven Water Pollution Control Authority. Daily samples are collected, and results are updated weekly. This facility serves 200,000 people in the area.[c] |

**TABLE 1-1** Continued

| Wastewater Surveillance Program | | Unique Aspects of the Program |
|---|---|---|
| **University Programs** | University of California, San Diego | Under this ongoing program that started in May 2020, 340 buildings are monitored for viral activity, and more than 200 wastewater samplers are situated across the campus.[d] |
| | University of Arizona, Tucson | The university first analyzed samples for SARS-CoV-2 from utilities across the country and then began analyzing samples collected on campus. The university has developed action levels for its campus wastewater surveillance program and used the wastewater data to prevent outbreaks. |

NOTE: Table includes both NWSS-funded community wastewater surveillance programs and privately funded programs.
[a] See https://storymaps.arcgis.com/stories/f7f5492486114da6b5d6fdc07f81aacf.
[b] See https://returntolearn.ucsd.edu/dashboard/index.html.
[c] See https://www.yalecovidwastewater.com.
[d] See http://wbe.stanford.edu.
SOURCE: Adapted from EPA (2021) unless otherwise noted.

is the first national-level wastewater disease surveillance system in the United States, and it coordinates with state-, tribal-, local-, and territorial-level health departments to design and integrate wastewater surveillance data to inform public health decisions. The mission of the NWSS is to

- offer technical assistance to public health departments and wastewater utilities implementing wastewater surveillance;
- coordinate a centralized and standardized data portal for tracking of disease presence across the country;
- establish working groups for health departments, public health laboratories, and wastewater utilities for knowledge sharing; and
- strengthen epidemiological and laboratory capacity for wastewater surveillance at health departments (Kirby et al., 2021).

### Implementation

The rapid expansion and coordination of wastewater surveillance across the United States was an emergency response to the COVID-19 pandemic. SARS-CoV-2 was first detected in the United States in January 2020, and several wastewater surveillance efforts were under way in the spring of 2020, with support from local and state funding, federal emergency response grants, nongovernmental organizations, and philanthropic partners. By September 2020, formal pilot wastewater surveillance sites were established in eight states as part of the NWSS. As of October 2022, the NWSS comprises more than 1,250 sampling sites, covering a population of more than 133 million individuals. In fiscal year (FY) 2022, CDC awarded funding to support wastewater surveillance programs across 42 states, 5 cities, and 10 tribes.[2] To supplement jurisdiction-led wastewater surveillance programs, the NWSS provides testing capacity for an additional 500 sites through a commercial testing contract.[3] The NWSS continues to expand to new sites (Kirby, 2022). A map of participating sites and the distribution of sites as of October 2022 is illustrated in Figure 1-2. As shown by the map, the geographic distribution of sites is generally clustered near major metropolitan areas with a paucity of sites across the southern and intermountain western United States. NWSS sites are based in municipal wastewater systems; communities and populations that are unsewered are only captured in a wastewater surveillance system to the extent that individuals commute to a monitored sewershed for work, school, or other activities.

Implementation at NWSS-participating sites depends upon three primary entities—the health department, local wastewater utility, and analytical laboratory—to collect, test, and analyze samples and interpret the data (see Figure 1-1). Typically, local wastewater utilities collect, store, and distribute samples, which are used only for the NWSS and have no water quality regulatory implications. A public health, commercial, or academic laboratory partner analyzes the samples, and the public health department interprets the data to identify trends regarding infection prevalence within a community, integrate the wastewater data with other surveillance data, and determine the appropriate public health response. The multidisciplinary nature of a national wastewater surveillance system requires extensive collaboration between and across health departments, testing laboratories, and wastewater utilities. Historically, there has been limited collaboration between public health and wastewater departments (Clason, 2022), and active relationships across agencies rarely existed (McClary-Gutierrez et al., 2021).

At the national level, CDC provides funding support for these systems. CDC initially provided $2.5 million to support eight pilot NWSS sites, funded through the 2020 Coronavirus Aid, Relief, and Economic Security (CARES) Act.[4] The number of sites was expanded through an additional $33 million provided through the Paycheck Protection Program. Initial funding for participating NWSS sites came from Epidemiology and Labora-

---

[2] See https://www.cdc.gov/healthywater/surveillance/wastewater-surveillance/progress/index.html and https://www.cdc.gov/budget/fact-sheets/covid-19/funding/index.html.

[3] See https://sam.gov/opp/c68491abc61e4f6392b14d1e1abaf7c7/view.

[4] *Coronavirus Aid, Relief, and Economic Security Act of 2020*, Public Law 116-136, 116th Cong., 2nd sess. (March 27, 2020).

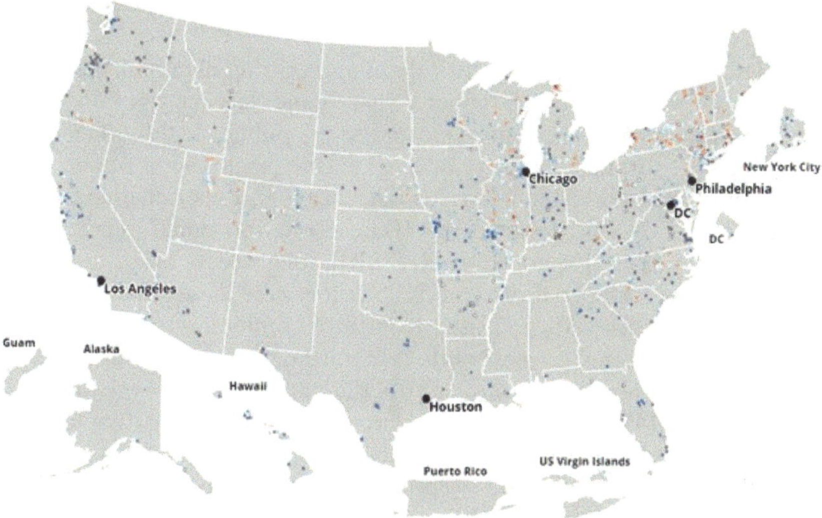


**FIGURE 1-2** National Wastewater Surveillance System dashboard as of October 2022.
SOURCE: https://covid.cdc.gov/covid-data-tracker/#wastewater-surveillance.

tory Capacity for Prevention and Control of Emerging Infectious Diseases (ELC) Cooperative Agreement grants provided by CDC to eligible health departments (i.e., state health departments, territories, and some large cities and counties).[5] An additional $200 million in grants were made available from the ELC Enhancing Detection and Enhancing Detection Expansion program, supported by the Coronavirus Response and Relief Supplemental Appropriations Act of 2021.[6] Finally, the NWSS was granted $384 million through the American Rescue Plan,[7] starting in FY 2022 for use through 2025. In FY 2022, the NWSS supported wastewater surveillance initiatives in 42 states and 5 cities, with a total of $64 million in funding. The average amount awarded to each jurisdiction in FY 2022 was approximately $900,000.[8] In addition, the NWSS has provided funding to 10 tribal communities to develop wastewater surveillance capacity through the Tribal

---

[5] All 50 states, 5 cities, 1 county, and 8 territories have been awarded ELC funding. See https://www.cdc.gov/ncezid/dpei/elc/elc-recipient-history.html.

[6] *Coronavirus Response and Relief Supplemental Appropriations Act of 2021*, Public Law 116-260, 116th Cong., 2nd sess. (December 27, 2020).

[7] *American Rescue Plan Act of 2021*, Public Law 117–2, 117th Cong., 1st sess. (March 11, 2021).

[8] See https://www.cdc.gov/budget/fact-sheets/covid-19/funding/index.html.

Public Health Capacity Building and Quality Improvement cooperative agreement.[9]

CDC also plays an important role in aggregating data and sharing the results from participating wastewater surveillance sites across the country. Data from the NWSS are communicated to the public, health officials, and policy makers through a variety of mechanisms, including a public-facing data dashboard, weekly summarized data briefs,[10] a restricted-access data dashboard for health departments (Data Collation and Integration for Public Health Event Response [DCIPHER]), and weekly briefs for federal policy makers.[11] The goal of the NWSS is for these data to be interpreted by public health officials and used to inform community health interventions, to raise public awareness of disease transmission within communities, and to track pathogen dynamics across the nation. CDC also coordinates Communities of Practice to build capacity among the participating localities and hosts monthly meetings with cohorts of participants across jurisdictions to share experiences and keep health officials apprised of updates or program improvements.

As part of the federal support for the NWSS, CDC and HHS also convene the National Sewage Surveillance Interagency Leadership (NSSIL) Committee, in which additional federal agencies collaborate and coordinate to exchange information and discuss agency-specific roles and activities.[12] CDC, U.S. Environmental Protection Agency, U.S. Department of Defense, U.S. Department of Homeland Security, U.S. Geological Survey (USGS), National Institutes of Health, and U.S. Department of Veterans Affairs support implementing sewage sampling and developing guidance documents for use by public health officials. CDC and USGS coordinate to provide surge capacity for wastewater testing when needed. Federal agencies also coordinate to prioritize federal research on wastewater sampling, analysis, and interpretation. Finally, NSSIL coordinates with several nongovernmental organizations, including the Association of Public Health Laboratories, the Association of State and Territorial Health Officials, the Water Environment Federation, and the Water Research Foundation (see Figure 1-3).

### Current Status and Future Outlook

As the COVID-19 pandemic continues with ongoing monitoring of emerging variants and subvariants and possibly transitions from emer-

[9] See https://www.cdc.gov/tribal/cooperative-agreements/tribal-capacity-building-OT18-1803.html.

[10] See https://www.cdc.gov/coronavirus/2019-ncov/covid-data/covidview/index.html.

[11] See https://covid.cdc.gov/covid-data-tracker/#wastewater-surveillance.

[12] See https://www.cdc.gov/healthywater/surveillance/wastewater-surveillance/federal-coordination-partnering-wastewater-surveillance.html.

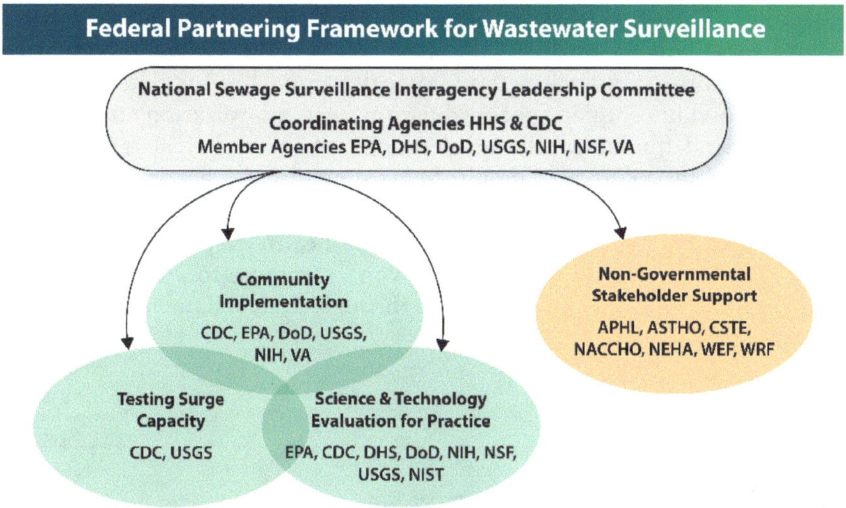


**FIGURE 1-3** National wastewater surveillance system federal partnering framework.
NOTES: APHL = Association of Public Health Laboratories; ASTHO = Association of State and Territorial Health Officials; CDC = U.S. Centers for Disease Control and Prevention; CSTE = Council of State and Territorial Epidemiologists; DHS = U.S. Department of Homeland Security; DoD = U.S. Department of Defense; EPA = U.S. Environmental Protection Agency; NACCHO = National Association of County and City Health Officials; NEHA = National Environmental Health Association; NIH = National Institutes of Health; NIST = National Institute of Standards and Technology; NSF = National Science Foundation; USGS = U.S. Geological Survey; VA = U.S. Department of Veterans Affairs; WEF = Water Environment Foundation; WRF = Water Research Foundation.
SOURCE: https://www.cdc.gov/healthywater/surveillance/wastewater-surveillance/federal-coordination-partnering-wastewater-surveillance.html.

gency response to endemic disease management, the application of wastewater surveillance as a public health tool will evolve. In particular, state, tribal, local, and territorial public health professionals; public utilities; and CDC are reviewing the usefulness of wastewater surveillance to inform public health decisions for SARS-CoV-2 as well as potential applications to other infectious pathogens. The surveillance system is also at a point of transition from an ad hoc collection of willing state and local participants seeking all useful information for local emergency pandemic response to a forward-looking national wastewater surveillance system that serves state, tribal, local, territorial, and national public health objectives simultaneously. Questions remain about what a standardized national wastewater surveillance system should look like, including ethical and privacy considerations; standard methodological approaches for data sampling, analysis, and interpretation; coordination or standardization among jurisdictions; and fundamental considerations of the technical feasibility of wastewater surveillance to monitor emerging diseases beyond COVID-19 in the United States. In addition, uncertainty remains around the use of wastewater surveillance to inform public health response, particularly how this form of disease monitoring can contribute to and complement traditional public health surveillance through clinical data and syndromic surveillance.

**BOX 1-4**
**Statement of Task**

**Phase 1**

An ad hoc committee of the National Academies of Sciences, Engineering, and Medicine will review community-level wastewater-based disease surveillance and its potential value toward prevention and control of infectious diseases in the United States. The committee will:

1. Describe wastewater-based disease surveillance and how it differs from other approaches to disease surveillance and other wastewater monitoring for contaminants.
2. Review how wastewater-based surveillance has been useful in understanding COVID-19 in communities and informing local public health decisions.
3. Examine the potential value of specific applications of wastewater-based disease surveillance for understanding and preventing disease and illness beyond COVID-19 and factors that may limit its application in the United States.
4. Describe the general characteristics of a robust, integrated approach for national use of wastewater-based disease surveillance.
5. Discuss broad approaches to increase the public health impact of wastewater surveillance in the United States and the most effective strategies for federal, state, and local coordination to achieve national implementation of wastewater surveillance for an array of diverse infectious disease health indicators.

For the purpose of this study, community-level wastewater-based disease surveillance implies sampling at wastewater treatment plants and does not include local surveillance at neighborhood or institutional scales. To inform the study, the committee will briefly review ongoing and planned U.S. federal, state, local, tribal, and territorial efforts as well as international case examples for implementing

## MOTIVATION FOR THE STUDY

CDC charged the National Academies of Sciences, Engineering, and Medicine to appoint a committee of academic and public health experts to review community-level wastewater surveillance and its potential value toward understanding and preventing infectious disease in the United States. The committee's work has been divided into two parts as described in Box 1-4. The first phase, which is the focus of this report, provides an assessment of the usefulness of current community-level wastewater surveillance in the United States and its potential value for infectious disease beyond COVID-19. As explained in the statement of task, in the context of this study, "community-level" wastewater surveillance includes "sampling

wastewater-based disease surveillance. The committee's report will include conclusions and recommendations on wastewater-based surveillance in federal, state and local public health efforts in the prevention and control of infectious diseases. Applications of wastewater-based surveillance for non-infectious agents, in global settings, and for facility-level surveillance are outside the scope of this review, but the committee may identify these for future evaluation.

**Phase 2**

The committee will conduct an in-depth study of opportunities and barriers relevant to increasing the use and utility of wastewater surveillance for the prevention and control of infectious diseases in the United States. Specifically, the committee will:

1. Define specific characteristics for development and implementation of a robust, integrated wastewater-based infectious disease surveillance program and discuss technical constraints and opportunities associated with wastewater sampling, testing, and data analysis, including:
    - Methods and/or quality criteria, including genomics and sequencing, to detect pathogens, including strain- or variant-specific methods. Methods for discovery of unknown emerging pathogens can also be considered.
    - Data reporting, data analysis, and data interpretation for detecting emerging threats to public health and estimating disease incidence and prevalence, including data integration with other surveillance data for improving predictive models.
2. Identify significant technical limitations that could impact the feasibility of using wastewater surveillance as a platform for generating data for indicators of public health status and risk.
3. Describe the research, development, and information sharing needed to meet emerging needs and increase impact of wastewater surveillance for improving public health in the United States.
4. Identify resources for supporting wastewater surveillance.

at wastewater treatment plants" and does *not* include "local surveillance at neighborhood or institutional scales." However, in committee discussions with the sponsor, hyperlocal sampling at specifically designated sentinel sites, such as likely points of entry of infectious disease, was deemed to be within the scope of the study because the intent of these sites is to provide data of value to the nation (see also Chapter 3). In addition, a few examples of sub-sewershed and institutional-scale surveillance are included in the report to accurately portray the range of wastewater surveillance efforts that took place during the COVID-19 pandemic, and to highlight lessons learned that may be applicable to community-scale efforts. The committee was not asked to assess non-infectious agents or surveillance in global settings. The Phase 2 study (see Box 1-4) will offer a detailed technical evaluation and needs assessment for an ongoing national wastewater surveillance program.

To address its Phase 1 statement of task, the committee held two information-gathering meetings. Speakers were selected to complement the broad and interdisciplinary experiences of the committee members, in particular representing perspectives from utility, public health, and ethics stakeholder groups engaged in wastewater surveillance. These discussions served as the initial basis for the committee's deliberations, which were further informed by a review of relevant literature and the committee's collective expertise.

## REPORT STRUCTURE

This report describes the usefulness of a robust community-level wastewater surveillance system for the United States and highlights approaches for integrating wastewater surveillance data into a public health response for a variety of pathogens. Chapter 2 provides a retrospective assessment of how wastewater surveillance was used in understanding and informing the public health response during the COVID-19 pandemic, including early challenges that were encountered. Chapter 3 describes a vision for a national wastewater disease surveillance system, including key characteristics of a robust system. Chapter 4 discusses strategies for implementing the committee's vision for a national wastewater-based infectious disease surveillance system beyond COVID-19, discussing future challenges and strategies to collaborate across federal, state, and local jurisdictions.

# 2

# Wastewater Surveillance for COVID-19

Wastewater infectious disease surveillance was implemented in many locations in the United States and globally during the COVID-19 pandemic and continues to be used to track ongoing disease outbreaks and the spread of variants. In this chapter, the committee reviews how wastewater surveillance has been useful in understanding COVID-19 in communities and in informing local public health decisions. Although the committee's task (and the National Wastewater Surveillance System [NWSS]) emphasizes community-level surveillance, in this chapter the committee also includes a few examples of institutional and sub-sewershed sampling (labeled as such) to demonstrate how information has been useful at different scales in ways that may inform the broader potential benefits of national wastewater surveillance.

## VALUE FOR UNDERSTANDING COVID-19 IN COMMUNITIES

Since the emergence of COVID-19 in early 2020, U.S. epidemiological surveillance has incorporated a number of conventional data sources to track COVID-19 burdens and trends, including clinical test results and case information, COVID-19 hospitalizations (compiled through the HHS [U.S. Department of Health and Human Services] Unified Hospital Data Analytic Dataset[1]), and COVID-19 deaths. Each of these data types have limitations that have hindered real-time understanding of community COVID-19 burdens and trends. Routine testing results have been regularly reported for

[1] See https://www.cdc.gov/coronavirus/2019-ncov/covid-data/covidview/index.html.

U.S. counties, usually as new cases per 100,000 inhabitants, but there are issues with this source of data, including the large costs of testing all suspected cases and the biases that come from changes in testing availability. Furthermore, home-based antigenic testing increased greatly in 2022, and positive results may not be reported to public health authorities (Ritchey et al., 2022). This has decreased the use of laboratory-based tests and exacerbated case underreporting (Rader et al., 2022). Although COVID-19 hospitalization and death data lack some of the biases associated with clinical test data, hospitalizations and deaths lag behind COVID-19 infections. Deaths from COVID-19, for example, have been shown to cluster approximately 17 to 21 days after infection (Ward and Johnsen, 2021). In addition to the inherent *time lags* of hospitalizations and deaths from infections that stem from the progression of the disease, each of these conventional data sources take time to reach public health agencies. This leads to *time delays* in posting and using the data. For example, the Washington State Department of Health requires 7 days to collect, quality check, and report hospitalization data.[2] Time delays differ across data sources and locations and can change over time for a given location.

Wastewater surveillance has been increasingly used to supplement these conventional data sources as it addresses some of the information gaps. Regardless of symptomatic status, a large fraction of individuals infected with SARS-CoV-2 shed virus through their stool (Zhang et al., 2021). Although people also shed SARS-CoV-2 in saliva, mucous, and urine, feces has been shown to be the dominant source into wastewater (Crank et al., 2022). Wastewater surveillance is a passive measurement, meaning it does not require the active participation of individuals in the healthcare or testing systems. As such, it avoids testing availability and behavior biases associated with clinical case data and is not affected by the increasing trend of at-home testing.

Once it was demonstrated that SARS-CoV-2 wastewater concentrations correlated with cases, questions were quickly raised about the potential of wastewater data to provide more timely information on the dynamics of COVID-19 in communities than case or hospitalization data. In other words, could wastewater data be a *leading indicator* of the traditional surveillance data time series and provide an early warning of clinical trends? If so, could they help direct more timely public health decisions? If rising concentration in wastewater during low-incidence periods was an early indicator of rising cases and hospitalizations, public health officials could make earlier recommendations for social distancing or masking and help hospital administrators decide when to cancel elective surgeries. Likewise, if wastewater concentrations peaked before case data or hospitalizations

---

[2] See https://doh.wa.gov/emergencies/covid-19/data-dashboard.

peaked, communities could make earlier decisions to scale back their emergency responses (e.g., opening additional COVID-19 units) and use the related resources for other purposes.

In the following sections, the committee reviews how wastewater surveillance has been useful in understanding COVID-19 data trends and spatial distribution in communities, the spread of variants, and the potential for early warning.

### Data Trends

Early work on wastewater surveillance of COVID-19 sought to demonstrate that wastewater concentrations correlated with case data collected through standard surveillance. Indeed, data from 2020 showed wastewater concentrations of SARS-CoV-2 correlating closely with case data once clinical testing was available (Ahmed et al., 2020; Graham et al., 2021; Medema et al., 2020; Peccia et al., 2020). In addition to case count data, wastewater concentrations correlated to clinical positivity rates (D'Aoust et al., 2021; Hopkins et al., 2022) and hospitalizations (D'Aoust et al., 2021; Peccia et al., 2020). The correlations between wastewater and epidemiological data were poor, however, very early in the pandemic when clinical testing was not routinely available (Graham et al., 2021). Rather than suggesting a problem with wastewater surveillance, this observation highlighted the ability of wastewater measurements to capture COVID-19 levels and dynamics in the absence of widespread access to and use of testing. Combined, these early reports demonstrated that wastewater measurements were, in fact, a valuable data source for COVID-19 surveillance.

Soon after the early reports showed correlations between wastewater levels and conventional surveillance data for a given locality, both internal and public-facing dashboards were developed to provide up-to-date SARS-CoV-2 wastewater levels. Numerous local dashboards have been developed to date, although some have stopped posting data as of summer 2022.[3] Some local dashboards present the wastewater data only (see Figure 2-1), whereas others present wastewater data alongside standard surveillance data (see Figure 2-2). In addition to public-facing dashboards, data summaries showing wastewater levels have been posted on internal dashboards and sent in regular emails to stakeholders (see Figure 2-3).

Data in published reports and on dashboards have demonstrated the value of understanding SARS-CoV-2 wastewater data trends in both high and low–infection prevalence settings. In high-prevalence scenarios, quantitative wastewater results identify increasing, plateauing, or declining trends in community infection prevalence. In low-prevalence scenarios, wastewater

[3] See, for example, https://covidwastewatermonitor.wyo.gov.

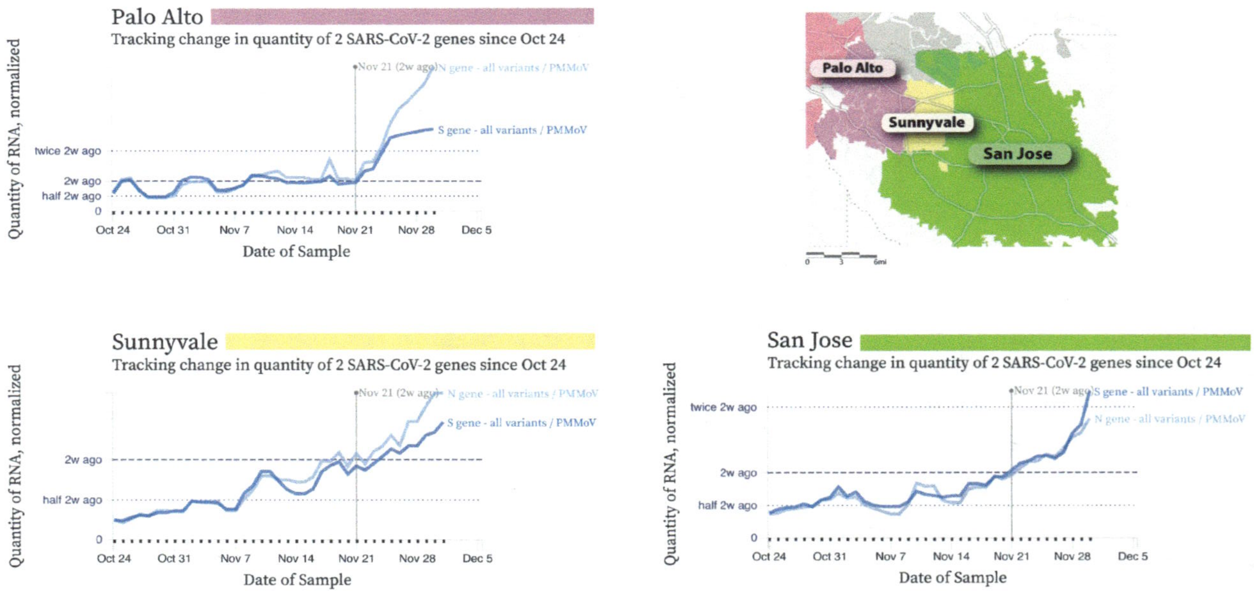


**FIGURE 2-1** Example dashboards that have presented SARS-CoV-2 wastewater data without other surveillance data from the Sewer Coronavirus Alert Network (SCAN) dashboard covering Northern California. The data are from three neighboring communities shown on the inset map: Palo Alto (purple), Sunnyvale (yellow), and San Jose (green).
NOTE: Data are normalized against concentrations of the pepper mild mottle virus (PMMoV), which is commonly found in pepper-based food items and in human feces, to account for dilution by other inflows (e.g., industrial, stormwater) and thereby reduce one source of variability in the data.
SOURCE: Sewer Coronavirus Alert Network, http://wbe.stanford.edu. Courtesy of Zan Armstrong, Alexandria Boehm, and Marlene Wolfe.

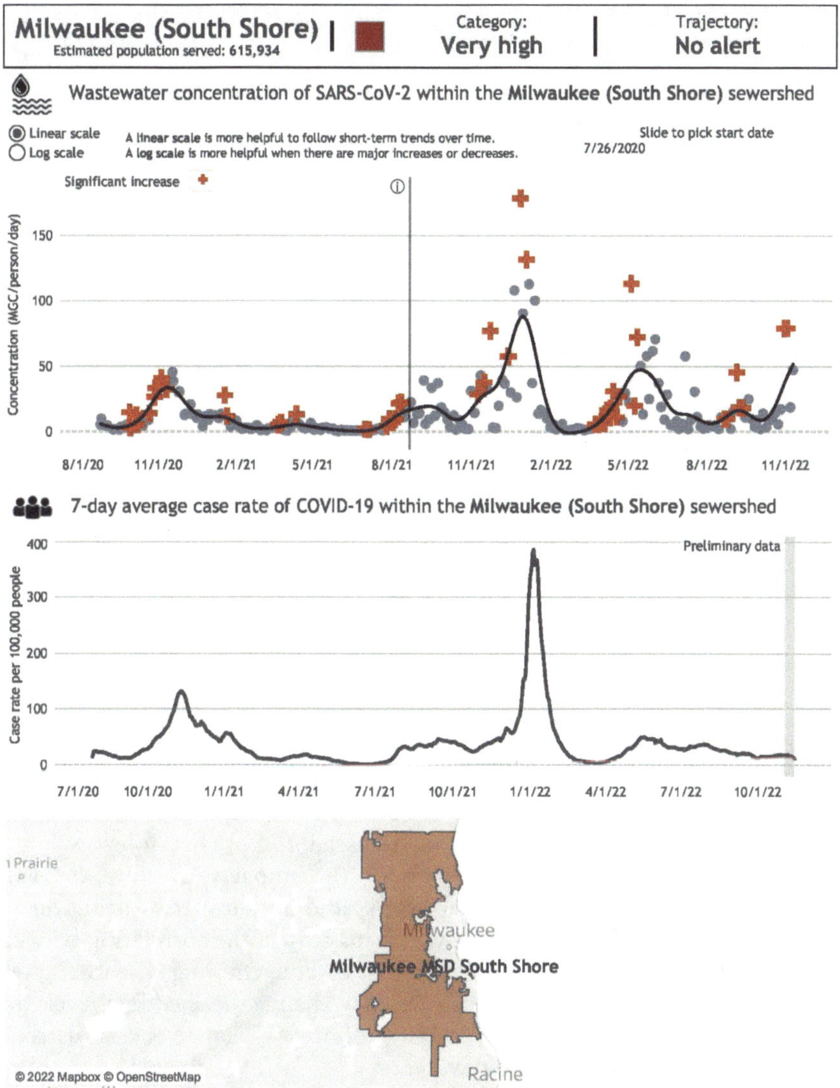


**FIGURE 2-2** Example dashboard that has presented SARS-CoV-2 data in addition to other surveillance data from Milwaukee, Wisconsin. This public dashboard includes a time series plot of SARS-CoV-2 concentration (as million gene copies [MCG]/person/day) in wastewater compared to polymerase chain reaction positive case count in Milwaukee Municipal Sanitation District South Shore, Wisconsin.
SOURCE: https://www.dhs.wisconsin.gov/covid-19/wastewater.htm.

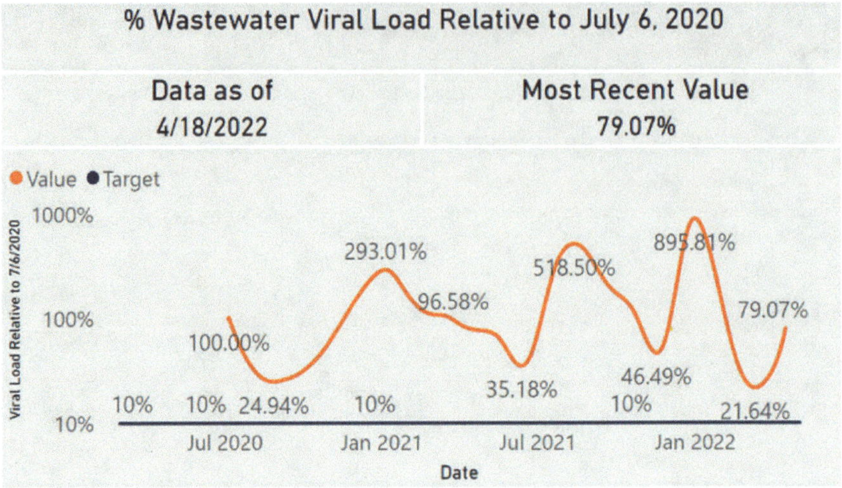


**FIGURE 2-3** Example of wastewater data posted on an internal Houston Health Department indicators and trends dashboard that is not viewable by the public. SOURCE: Houston Health Department.

results can help identify when the pathogen begins to circulate in the community or when prevalence starts to increase. The trend data are arguably most valuable when used in combination with other data sources (e.g., to confirm observations from clinical data and investigate abnormal trends). In some cases, wastewater trends are seen before other COVID-19 health metrics (e.g., Peccia et al., 2020), as discussed in more detail below.

Important questions remain about the impacts of different variants, prior infections, vaccination status, and antiviral treatments on the dynamics observed in wastewater, which may affect the correlation between viral load in wastewater and the results of conventional disease surveillance. Vaccination status, for example, may change the shedding levels and dynamics with infection, which would ultimately change the viral trends observed in a sewershed's wastewater.

There is an interest in using wastewater to estimate the overall COVID-19 disease burden in communities rather than simply using correlations to predict incidence or hospitalization trends (McMahan et al., 2021; Soller et al., 2022). To date, there are limited studies on fecal shedding, especially over time and with different variants, demographics, and vaccine statuses. Ongoing stool shedding studies are critical to help clarify how these variables affect the wastewater trends observed in communities and whether reliable and accurate prevalence predictions are possible with wastewater.

## Early Warning

When discussing wastewater data as a potential early warning or leading indicator compared to conventional surveillance data, it is important to understand that the wastewater data could lead conventional surveillance data for two reasons. First, people may excrete the virus in stool before they show symptoms and seek testing. At the time of this writing, there were limited data sets on fecal shedding early in infections (e.g., prior to symptom onset); as a result, there is insufficient mechanistic evidence to establish that measurement of SARS-CoV-2 RNA in wastewater leads clinical testing or hospitalizations by a defined number of days. Second, there may be shorter time delays for wastewater analysis and communication of results to reach public health officials than for analysis and communication of results for clinical tests and reporting of hospitalization data. Wastewater data represent a single measurement per day for a whole sewershed, whereas case data represent many measurements, with samples collected in different clinics and analysis performed in different laboratories with different tests. A hypothetical timeline with the major surveillance steps is presented in Figure 2-4.

Multiple studies have retroactively evaluated the temporal relationship between increases in SARS-CoV-2 RNA in wastewater and reported

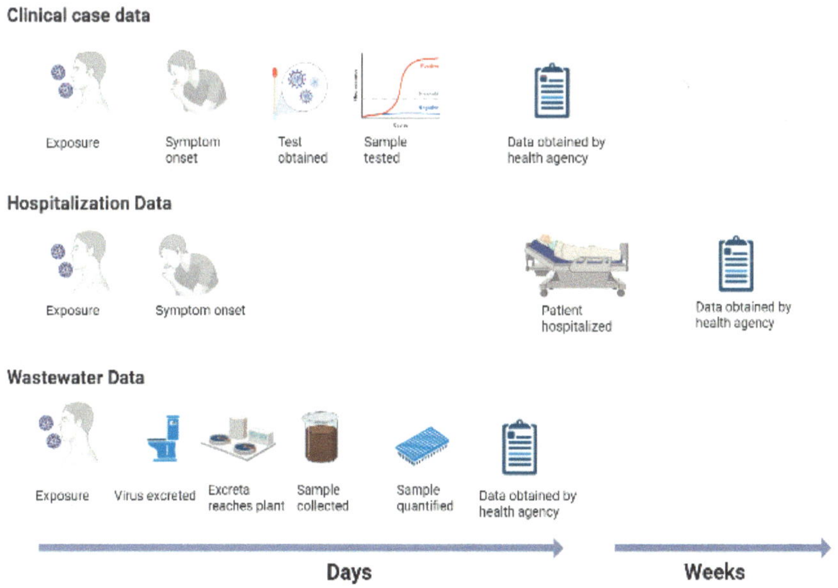


**FIGURE 2-4** Steps in the reporting of conventional and wastewater surveillance data during the COVID-19 pandemic.
SOURCE: Created with BioRender.

clinical cases (D'Aoust et al., 2021; Fernandez-Cassi et al., 2021; Galani et al., 2022; Peccia et al., 2020; Wu et al., 2022a), and the findings have varied. For example, Wu et al. (2022a) compared retrospective wastewater concentrations (sampling date) from early 2020 with reported cases (testing date) and reported that wastewater trends preceded case data trends by 4–10 days. By contrast, Peccia et al. (2020) reported no wastewater lead time over the same time frame in the COVID-19 pandemic when comparing the wastewater sampling date with clinical specimen collection dates. Feng et al. (2021) also reported no lead time for wastewater surveillance data when sample and clinical specimen collection dates were compared over 12 sewersheds. The different conclusions regarding lead times reported between studies could stem from numerous factors, including

- the phase of the pandemic, with different phases having different testing availability and potentially changing viral shedding characteristics associated with different variants and shifting natural and vaccine-induced immunities (Wu et al., 2022b);
- different amounts of time required for sample transport, analysis, and data reporting;
- different wastewater sampling frequencies;
- the sewershed structure and population; and
- the specifics of the conventional surveillance data to which the wastewater data are compared (e.g., date of specimen collection, date of symptom onset, and date of data reporting).

SARS-CoV-2 wastewater data have the potential to be reported more quickly or along a more consistent time frame compared to conventional surveillance reporting (see Figure 2-4). Indeed, in contrast to what they observed when comparing data by wastewater sampling date and clinical specimen collection date, Peccia et al. (2020) observed a 6- to 8-day lead time in wastewater trends when they compared wastewater sampling date to the date that cases were reported to public health officials. This demonstrates the importance of including reporting time delays when assessing if wastewater data trends lead other surveillance data trends.

Overall, the extent of the wastewater SARS-CoV-2 RNA concentration trends leading clinical case data trends changed over the course of the pandemic (Xiao et al., 2022). Wastewater dynamics data exhibited greater lead times when tests were scarce and when case data reporting to public health agencies was delayed. Whereas studies often consider or discuss case reporting delays in their analyses, they rarely discuss wastewater reporting delays. As with case data, it can take anywhere from a single day to weeks for wastewater samples to be measured and the data to be posted to dashboards or shared with health officials. For accurate assessments of wastewater lead times, it is important to compare the reporting date of both clinical case counts and wastewater concentrations for public health officials.

Whereas the results of wastewater surveillance data leading clinical case data are mixed, wastewater data trends have consistently led hospitalization and death data trends regardless of the phase of the pandemic or the location of the study (D'Aoust et al., 2021; Galani et al., 2022; Peccia et al., 2020). Peccia et al. (2020) reported a 1- to 4-day lead in wastewater data compared to hospitalizations. Similar lead times were reported by Galani et al. (2022), with a mean of 5 days, atlhough Saguti et al. (2021) noted lead times of 19–21 days for wastewater data compared to hospitalization in early 2020. Note that these conclusions did not take reporting delays of either wastewater measurements or hospitalization counts into account. Hospitalization and death reporting delays would result in greater lead times for wastewater surveillance data.

In summary, the upward and downward trends in the levels of wastewater SARS-CoV-2 RNA may not always lead the case data trends when the dates of wastewater sampling and dates of testing are compared. Trends in wastewater SARS-CoV-2 RNA do consistently lead hospitalization trends when dates of wastewater sampling and hospitalization are compared, although with reporting delays, the lead time may be reduced to only a modest amount (1–5 days). The value of wastewater surveillance data as a leading indicator of conventional surveillance data can be enhanced at times when there are significant testing delays and case/hospitalization/death number reporting delays *and* when wastewater data *do not* have a significant reporting delay.

## Localized Spatial Information

Beyond its use in identifying trends and burdens within a community, wastewater surveillance has been used to compare COVID-19 trends between communities. Throughout COVID-19, numerous organizations and agencies have monitored wastewater in multiple areas served by wastewater treatment facilities (i.e., sewersheds) within a region. The early implementation of wastewater monitoring took place in locations that had the expertise and resources to implement the technology. As such, the sites that participated across the United States were not evenly spaced (see Figure 1-2), and monitoring was not designed to be representative of the U.S. population. Nonetheless, the results from different locations provided insight on the spatial variation in COVID-19 trends and burdens.

Some large cities and counties are served by several wastewater treatment plants and therefore are made up of several sewersheds, allowing more localized spatial information in trend monitoring of SARS-CoV-2

RNA from wastewater treatment plant sampling. For example, to identify localized trends, Houston, Texas collected and analyzed samples from 39 wastewater treatment plants throughout the metropolitan area that collectively serve more than 2 million people (see Figure 2-5). Often, sewersheds within the same U.S. regions exhibited similar SARS-CoV-2 dynamics, but wastewater data can demonstrate when COVID-19 levels are deviating from nearby communities. For example, both Ann Arbor and Ypsilanti are within Washtenaw County, Michigan. The COVID-19 case data are typically aggregated for the entire county on public health dashboards. Wastewater data from the two communities, which are analyzed with the same methods in the same laboratory, have provided higher spatial resolution compared to the aggregated county case data. The wastewater data have identified a number of instances in which the two communities in the same county had opposite trends, as well as different relationships between wastewater levels and reported cases (e.g., Figure 2-6).

Some sewersheds are very large, serving more than 1 million residents. Several studies sampled at locations within a sewershed (e.g., lift stations)

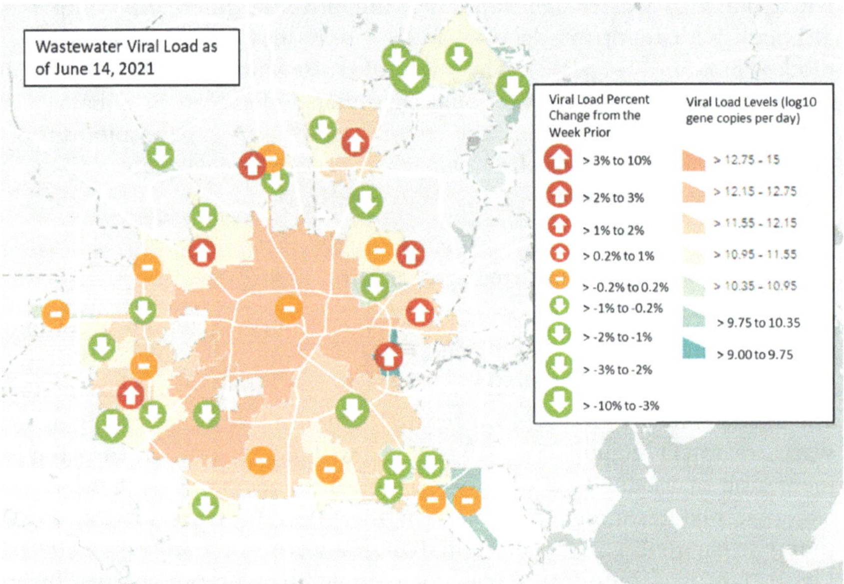


**FIGURE 2-5** Spatial plot of 39 wastewater treatment plants in Houston, Texas, colored by viral load as of June 14, 2021, and with arrows indicating whether the location was experiencing an increasing, decreasing, or plateauing trend compared to the previous week. The size of the arrow signifies the magnitude of the trend. SOURCE: https://covidwwtp.spatialstudieslab.org.

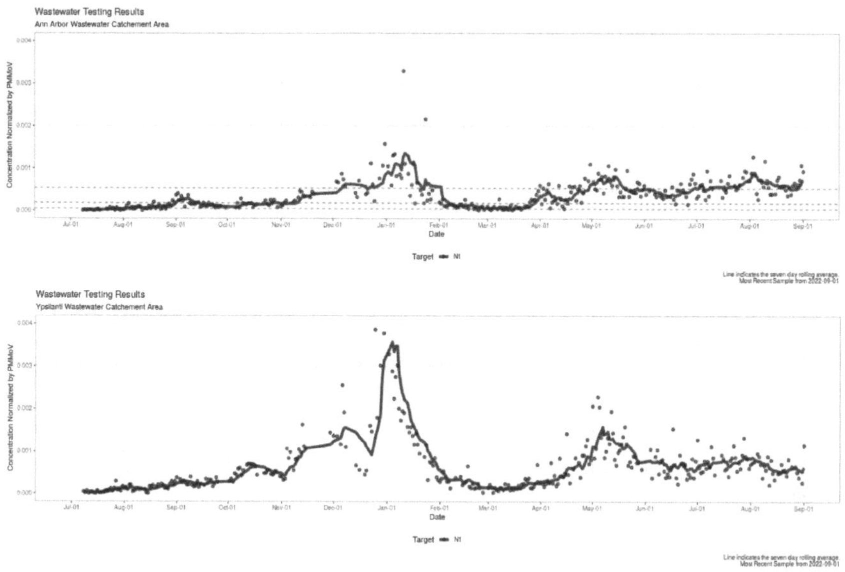


**FIGURE 2-6** Normalized wastewater concentrations of SARS-CoV-2 RNA from Ann Arbor, Michigan (top) and Ypsilanti, Michigan (bottom) from July 2021 through August 2022. The two communities are located in the same county, and case data from the county are typically aggregated on public-facing dashboards. In this case, the community-scale wastewater data provide higher spatial resolution than that provided by aggregated county surveillance data and have identified differences in trends and burdens between the two communities.
NOTE: PMMoV = pepper mild mottle virus.
SOURCE: https://um.wastewatermonitoring.dataepi.org.

in addition to sampling at the central wastewater treatment plant, which averages trends in viral shedding across the entire sewershed (see Box 1-3). This can identify trends in sub-sewershed data that were not detectable in sewershed-level observations. For example, Layton et al. (2022) divided Newport, Oregon, communities into 22 regions for sub-sewershed wastewater surveillance; the results were combined with other data to identify "hot spots" (see Figure 2-7).

Using wastewater surveillance at a subregional scale can provide information about the spread of disease in areas with reduced availability of clinical testing to improve health equity. Many of the early wastewater surveillance sites focused on college towns, campuses, and urban areas with populations that are not representative of the U.S. population. In their study focused on Jefferson County, Kentucky, Yeager et al. (2021) proposed an approach to site selection that includes overselecting geographic areas

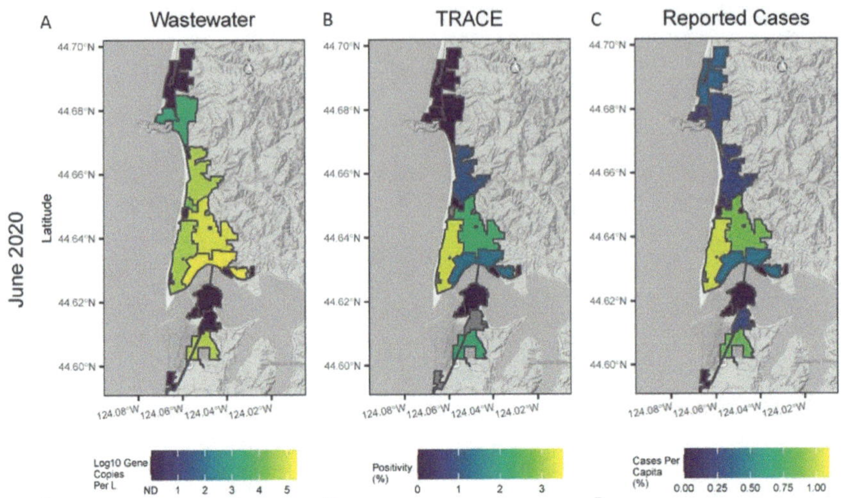


**FIGURE 2-7** Heatmaps of COVID-19 burden and SARS-CoV-2 wastewater data from 22 sub-sewersheds in Newport, Oregon. Data were used to identify "hot spots." Three data sets are compared here, including sub-sewershed wastewater surveillance from samples collected from pump stations, random door-to-door clinical sampling via nasal swabs by the Team-based Rapid Assessment of community-level Coronavirus Epidemics (TRACE) project, and reported cases per capita.
SOURCE: Layton et al. (2022).

that are "often underrepresented in clinical testing, such as low-income neighborhoods and communities of color."

Comparative analysis of SARS-CoV-2 RNA levels in wastewater across communities is complicated by several factors. Different laboratories and analytical methods can lead to differences in reported SARS-CoV-2 concentrations by greater than an order of magnitude (Islam et al., 2022; Pecson et al., 2021). Different sampling approaches, such as grab versus composite sampling and influent samples versus settled solids sampling, can also affect the SARS-CoV-2 signals. Additionally, variations in the characteristics of a community's wastewater and wastewater collection system can affect measured SARS-CoV-2 concentrations. Wastewater strength, or the concentration of human fecal matter in the wastewater, is particularly important because the feces are the greatest sources of SARS-CoV-2 in wastewater (Crank et al., 2022). Wastewater strength can vary between sewersheds within a region based on factors such as stormwater contributions to the wastewater, infiltration, the per capita water usage of the community, and various industrial inputs to the wastewater. To address these sources of variation, many groups conducting COVID-19 surveillance have incorporated normalizing factors (e.g., wastewater flow rates, suspended solids, pepper mild mottle virus [PMMoV], cross-assembly phage [crAssphage]) in their SARS-CoV-2 reporting to facilitate direct comparison of COVID-19 burdens between communities. Normalizing factors are frequently used to adjust for the fecal strength of wastewater. Additional factors that can complicate cross-sample comparisons are the wide ranges in sewershed sizes, which affect the amount of time wastewater spends in the sewage system and thus the degradation of the SARS-CoV-2 signal.

## Variants

Mutations in the SARS-CoV-2 genome have led to new variants throughout the course of the COVID-19 pandemic, and some variants have caused increased transmissibility, increased disease severity, and increased ability to escape immunity induced by either vaccination or prior infection, and, as a result, decreased effectiveness of public health measures. Wastewater variant tracking has been used to complement clinical testing for early detection and understanding of the spread of known and emerging variants. Improved understanding of shifts in virus epidemiology can be useful for predicting the risk of a new surge, directing resource allocation and prioritization, and refining public health messaging (Kirby et al., 2021). Two approaches are used to identify variants in wastewater—polymerase chain reaction (PCR) and genome sequencing—each of which has specific advantages.

PCR-based testing for Alpha, Delta, and Omicron variants has become widely implemented in wastewater surveillance efforts (Kirby et al., 2022; Lee et al., 2021; Schussman et al., 2022; Yu et al., 2022) because it provides a more comprehensive representation of circulating variant diversity compared to the small fraction of clinical cases being analyzed for variants (Kantor, 2022). PCR assays are designed based on known variant genome sequences, which are usually determined through sequencing of clinical samples. Although a PCR assay takes time to design, optimize, and validate after an emerging variant is identified, once developed, PCR test results can be generated within hours, producing quantitative data on the relative amounts of variants circulating among the population in a sewershed (see Figure 2-8).

In contrast, sequencing approaches allow for the tracking of emerging variants, including identification of novel lineages, without prior knowledge of the suspected variant sequence (i.e., new variants can be identified de novo). Like PCR-based testing, sequencing wastewater is particularly useful in areas that have low clinical sequencing coverage in the population to determine whether an emerging variant has spread to that area (Crits-Christoph et al., 2021; Karthikeyan et al., 2022; Schussman et al.,

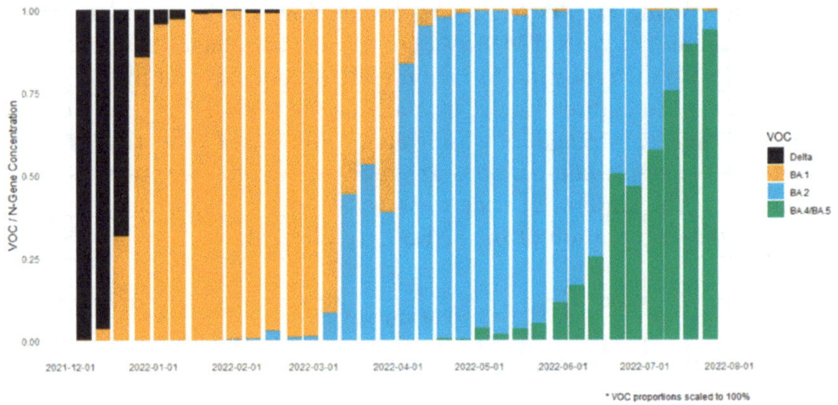


**FIGURE 2-8** Omicron subvariant tracking by droplet digital polymerase chain reaction. Results represent the proportion of each variant relative to N-gene concentration and are from eight mid-sized wastewater treatment plants in southeast Virginia.
NOTE: VOC = variant of concern.
SOURCE: Hampton Roads Sanitation District.

2022). Sequencing SARS-CoV-2 from wastewater provides substantial information (see Figure 2-9), although sensitivity for detecting low levels is reduced compared to PCR-based methods (Polo et al., 2020). The actual reconstruction of individual genomes is difficult due to the mixture of community lineages and is also time consuming. However, algorithms have been developed to infer the presence and relative abundance of different variants based on key mutations (Grubaugh et al., 2019).

Because of the complexity of the analysis, sequencing has not been performed as frequently as PCR-based testing in some cases. As an example, wastewater from all 14 wastewater treatment plants in New York City has been sequenced twice per month for the presence of novel variants, whereas PCR-based wastewater testing has been conducted weekly since June 2020 (Smyth et al., 2022). Technologies and analysis methods to sequence wastewater have matured substantially since the beginning of the COVID-19 pandemic, and more standardized methods have allowed more laboratories to be able to sequence wastewater samples. While the time from sample collection to results is still longer than a standard PCR test, the time gap between the two is closing.

Wastewater surveillance has in some cases provided information on variants prior to clinical data. For example, in the California Bay Area in late 2021, the Omicron variant was detected with PCR assays in some sewersheds before Omicron cases in the community were detected clini-

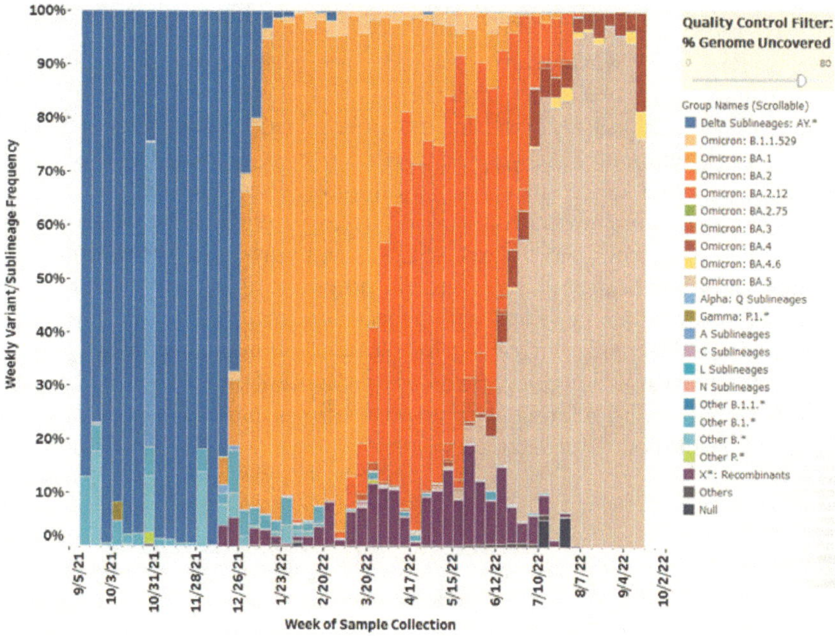


**FIGURE 2-9** Genome sequencing of SARS-CoV-2 variants and sublineages in wastewater from 12 participating sites.
SOURCE: U.S. Food and Drug Administration GenomeTrakr: https://www.fda.gov/food/whole-genome-sequencing-wgs-program/wastewater-surveillance-sars-cov-2-variants.

cally (Boehm and Riley, 2022). In this case, the quantitative PCR assay had been rapidly developed and implemented to capture the introduction of the Omicron variant to the sewershed. Karthikeyan et al. (2022) demonstrated that variants of concern could be identified in wastewater via high-resolution genome sequencing up to 2 weeks prior to their detection in clinical samples.[4]

## USE IN INFORMING PUBLIC HEALTH ACTIONS

In this section, the committee discusses the usefulness of wastewater surveillance data to inform public health decision making and to enhance public awareness of COVID-19 disease transmission, thereby informing public health–related actions.

[4] This retrospective study compared timing based on wastewater sampling date and clinical specimen collection data.

### Informing Agency Decision Making

SARS-CoV-2 wastewater data have been used extensively alongside other data sources to better understand the COVID-19 burden in a community in space and time and inform public health actions related to the virus in a community (see Figure 2-10; Kirby et al., 2021). When compared with other sources of information about disease burden, wastewater data can help inform decision making across a wide range of geographical scales (e.g., countries, states, cities, neighborhoods, institutions) and potential public health actions (e.g., warning, testing, vaccination, resource allocation, recommendations and orders concerning masking and social distancing). Although this report focuses on community-scale wastewater surveillance, institutional-scale wastewater surveillance was widely used to guide public health actions at those institutions, particularly colleges and prisons (see Box 2-1 for examples).

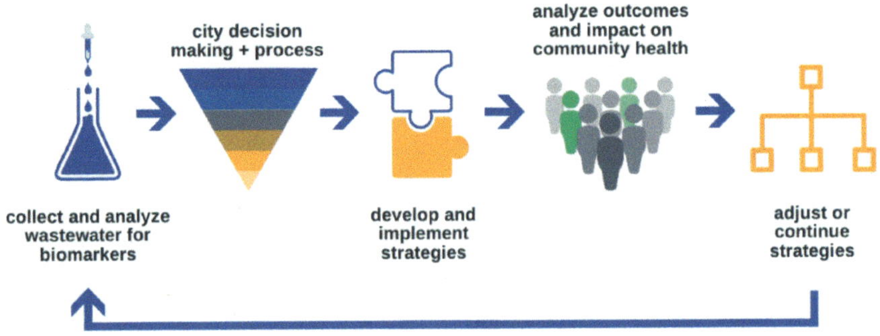


**FIGURE 2-10** Use of wastewater surveillance to advance public health strategies. SOURCE: https://covid19.tempe.gov.

**BOX 2-1**
**Institutional-Level Wastewater Surveillance for Public Health Management**

Wastewater surveillance proved promising at smaller sub-sewershed or institutional levels to prevent outbreaks. Many universities and correctional facilities used this tool to trigger individual testing efforts to identify infected individuals and isolate those who tested positive, which helped in preventing an outbreak and saved resources. Wastewater surveillance provided advanced warning to public health officials, providing a lead time to take measures and control a potential outbreak.

## BOX 2-1 Continued

For example, in the earliest stages of the pandemic, the Virginia Department of Corrections (VADOC) monitored SARS-CoV-2 RNA levels in wastewater from week to week and used the data to conduct focused monitoring to prevent widespread outbreaks of COVID-19 (M. Mayfield, VADOC, personal communication, 2022). At the University of Arizona, wastewater surveillance at dormitories identified the occurrence of asymptomatic and presymptomatic infection prior to clinical testing, leading to isolation and helping to avoid an outbreak. The University of Arizona was also able to track the effectiveness of interventions like the stay-at-home policy using wastewater surveillance by foreshadowing a decrease in SARS-CoV-2 RNA levels with a decrease in cases in the subsequent week (Kaiser, 2020). Wastewater surveillance was an important part of the University of California, San Diego's return to in-person learning. To help avert potential outbreaks, a large-scale wastewater surveillance system with building-level notifications was implemented covering 239 campus buildings (see Figure 2-1-1). Clinical testing of dorm residents based on the wastewater results enabled early diagnosis of 85 percent of COVID-19 cases (Karthikeyan et al., 2021).

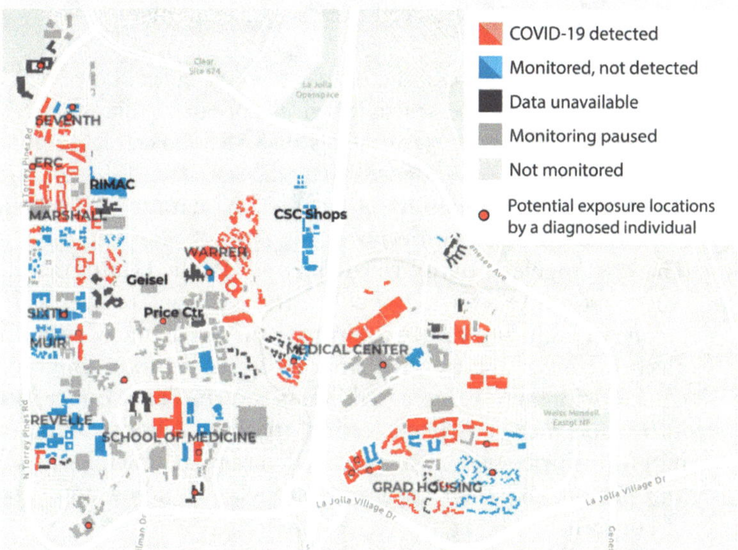


**FIGURE 2-1-1** Snapshot of the University of California, San Diego, COVID-19 dashboard, reflecting building-scale wastewater surveillance.
SOURCE: https://returntolearn.ucsd.edu/dashboard/index.html.

Wastewater surveillance data have been used to inform agency decision making and support a variety of public health actions at national, state, local, and institutional scales. Four key areas where these data have been particularly valuable include:

1. Confirming trends through comparison with other public health surveillance data,
2. Informing masking, social distancing, and stay-at-home policies,
3. Informing public health resource allocations, and
4. Informing clinical resource allocations.

### *Confirming Trends through Comparison with Other Public Health Surveillance Data*

When analyzed in comparison with other disease surveillance information, wastewater surveillance data are particularly useful in informing public health responses. By using multiple metrics in support of decision making, other sources of data are available to investigate anomalous results and increase confidence in observed trends. For example:

- In July 2020, Utah conducted wastewater surveillance across the state to examine the status of viral spread as diagnostic testing decreased. The decrease in levels of SARS-CoV-2 RNA in wastewater provided confidence that disease trends were actually decreasing, consistent with clinical testing data trends (Hopkins et al., 2022; McClary-Gutierrez et al., 2021).
- The Los Angeles County Department of Public Health includes the wastewater SARS-CoV-2 concentration relative to the peak value as one of eight indicators to provide the department with an early alert of concerning trends that could result in future high rates of transmission and/or increased illness severity. The county institutes mitigation measures (e.g., testing, masking indoors, contact tracing) for priority sectors such as skilled nursing units, K–12 schools, and persons experiencing homelessness, depending upon the level of concern.[5]
- In North Carolina, SARS-CoV-2 wastewater data are one of seven key metrics to monitor for increasing community spread and illness.[6] The weekly average SARS-CoV-2 virus copies found per capita from the 25 network sampling sites across the state are evaluated along with the percent of emergency room visits for

[5] See http://www.publichealth.lacounty.gov/media/Coronavirus/data/response-plan.htm.

[6] See https://covid19.ncdhhs.gov/dashboard.

COVID-19-like symptoms, the number of COVID-19 hospital admissions, the number of reported COVID-19 cases, vaccination and booster rates, the presence of COVID-19 variants among new cases, and CDC's COVID-19 community levels by county. All seven metrics are assessed together and compared over time to understand the current threat of the virus in the state.

### *Informing Masking, Social Distancing, and Stay-at-Home Policies*

Together with other disease surveillance data, wastewater surveillance has been cited as a valuable source of information to inform policies regarding social distancing, masking, and lockdowns:

- New Zealand's Prime Minister Jacinda Arden ended a 3-day lockdown for Auckland based on negative results for wastewater testing despite three new cases of COVID-19 in the county's largest city. She attributed her decision to community transmission data from wastewater surveillance and announced that the rest of the county would drop to Level 1 restrictions (de Jong, 2021).
- Using wastewater data combined with other surveillance data, the City of Tempe, Arizona, canceled public special events and meetings on March 19, 2020, and the mayor issued an order for the temporary closure of dining, fitness, and recreation venues.

### *Informing Public Health Resource Allocation*

Numerous examples show the value of wastewater surveillance data to inform decisions on the allocation of public health resources, such as testing sites, contact tracing, vaccination centers, and public notification efforts:

- In Ohio, a 10-fold increase in SARS-CoV-2 RNA levels within the state's wastewater surveillance system triggered an email warning to state and local health departments, utilities, and community leaders. The notifications were used to inform actions and allocate resources (Kirby et al., 2021). Within the warning areas, the state offered local health departments testing, vaccination, and contact tracing unit support. As of May 2022, 1,500 email warnings had been generated (Rebecca Fugitt, Ohio Department of Health, personal communication, 2022).
- In Utah, SARS-CoV-2 RNA levels in wastewater were incorporated in a metric alongside a 7-day positivity rate and 14-day case rates (equally weighted) to rank the state's 99 small area statistical health units for intervention. Decision makers used this risk rank-

ing, along with other considerations such as the existing resources in the area (e.g., existing testing sites), at a weekly meeting to prioritize where to send resources. An estimated 1.5 million tests were allocated based on this prioritization (Nathan LaCross, Utah, personal communication, 2022).

- In Oklahoma City, ZIP codes with high levels of SARS-CoV-2 RNA in wastewater relative to the rest of the city were identified. The Oklahoma City Public School District then conducted targeted communication to parents of students in high-load areas. Text messages encouraged parents to get tested and vaccinated and provided information on where to get tested and vaccinated. To complement the communication campaign, area hospitals and the health department deployed targeted vaccination and testing efforts in the area (Haley Reeves, University of Oklahoma, personal communication, 2022).
- Through a partnership with the Chesapeake Local Health Department, City Waterworks, Hampton Road Sanitation District, and local university partners, the State of Virginia used sub-sewershed-level wastewater sampling in the Chesapeake area to target vaccination campaigns to high-prevalence areas and educate the public (Cynthia Jackson, Chesapeake Health Department, personal communication, 2022).
- The City of Davis, California, used wastewater surveillance data to determine where to send geotargeted public health communications (e.g., door hangers, texts, phone calls, emails) on how to reduce the risk of contracting COVID-19 and where to get tested.[7] For example, in July 2021, 3,000 door hangers (see Figure 2-11) were delivered in response to spikes in levels of SARS-CoV-2 RNA in wastewater in three different neighborhoods.
- In June 2021, the Town of Beaufort, North Carolina, and the state health department issued a joint press release warning that elevated levels of SARS-CoV-2 were detected in the wastewater, although the number of new COVID-19 cases detected had not increased in testing results. Residents of the town were reminded to stay vigilant and get vaccinated (CCHD and NCDHHS, 2021; Virginia Guidry, North Carolina Department of Health and Human Services, personal communication, 2022).

[7] See https://healthydavistogether.org/the-new-pandemic-landscape-and-the-value-of-wastewater-monitoring/.

**FIGURE 2-11** Door hanger delivered in response to a spike of SARS-CoV-2 RNA in wastewater in the area.
SOURCE: https://healthydavistogether.org/the-new-pandemic-landscape-and-the-value-of-wastewater-monitoring.

### *Informing Clinical Resource Allocations*

A different example of wastewater data use was in the allocation of clinical resources. For example:

- In central Oklahoma, physicians used the breakdown of the Omicron and Delta variants in the wastewater to decide where to allocate monoclonal antibody therapies (Haley Reeves, University of Oklahoma, personal communication, 2022).
- In Houston, Memorial Hermann Health System found the city's data on levels of SARS-CoV-2 RNA in wastewater to be the most reliable indicator in predicting upcoming hospital capacity needs related to COVID-19. It found that trends of increasing SARS-CoV-2 RNA in wastewater provided an early indicator of an increasing rate of change of hospitalization by 2 weeks and used this information to prepare in the fall of 2021 and in the winter of 2022 when the hospital system's capacity was stressed. Decisions such as which unit to open next (e.g., post anesthesia care unit or common space), how to ensure adequate staffing for the new unit (e.g., contracts with nursing agencies or rescheduling provider staff), and whether to initiate deferrals of elective surgeries were based on the wastewater trends (Jamie McCarthy, Memorial Hermann Health System, personal communication, 2022).

### *Challenges Affecting the Use of Wastewater Surveillance Data*

There are multiple challenges that prevent local governments from fully adopting wastewater surveillance or fully putting the data to use. According to a survey of local and state governments, either competing priorities or lack of internal capacity was the leading reason for lack of adoption, affecting 72 percent of state and 47 percent of local agencies. Other factors included no clear designation of an agency responsible for wastewater surveillance, lack of funding, lack of buy-in from community leaders, and lack of partnerships with wastewater laboratories or utilities. About 21 percent reported that they did not understand how to interpret the data, and another 17 percent said the wastewater data did not add value to already existing information (Keshaviah et al., 2022b). In addition to the reasons listed, political considerations also play a role in limiting the impact of wastewater surveillance data, and sometimes politicians are not interested in wastewater findings. When wastewater analyses found Omicron to be dominant in Orange County, Florida, before clinical cases were present, Florida's hands-off policy meant that there was no change to public health actions (Vogel, 2022).

### Increasing General Public Awareness

The COVID-19 pandemic has led to a rapid increase in public access to public health metrics, such as numbers of reported cases, vaccination rates, and intensive care unit bed availability (Dixon et al., 2022). This public access has been enabled largely through the widespread adoption by public health agencies, news outlets, and academic research centers of online dashboards that are regularly updated with aggregated data in a straightforward and often interactive format. The public has shown strong interest in wastewater surveillance data. Wisconsin's dashboard, for example, was visited 127,000 times in the first 18 months (see Figure 2-12a). Likewise, Houston's dashboard was visited 107,000 times as of May 15, 2022, with as many as 10,000 clicks in a single day (see Figure 2-12b). Both dashboards experienced increased attention during COVID-19 surges. Similarly, the wastewater surveillance page on the U.S. Centers for Disease Control and Prevention's (CDC's) COVID Data Tracker has had more than 1 million views, and its public wastewater data have been downloaded more than 50,600 times between February 2, 2022, and July 13, 2022 (Rachel West, CDC, personal communication, 2022).

Early information and anecdotal accounts suggest that mentions of wastewater data by public health officials likely increase public awareness. For example, in the summer of 2020, Oklahoma wastewater surveillance indicated high levels of SARS-CoV-2 RNA in Anadarko, Oklahoma. The wastewater surveillance team informed the city manager who subsequently posted an alert on the city's Facebook page. The specific post was interacted with more than 6,000 times, in a town of 6,600 residents, in the following few days (Jason Vogel, University of Oklahoma, personal communication, 2022). In a similar example, the University of Louisville launched a geographically targeted awareness and action ad campaign in high-risk areas based on its wastewater surveillance data across the city. Ad campaigns were developed closely with community partners, using trends in wastewater surveillance data to increase vaccination. More than 2,500 people interacted with the ads, which reflects a click-through rate of 0.15 percent, compared to an industry-wide rate of 0.06 percent. The ads were used to inform the community when SARS-CoV-2 RNA levels in wastewater were both high and low (see Figure 2-13; Ted Smith, University of Louisville, personal communication, 2022). Increased risk awareness of COVID-19 trends is one of several factors that predicts increased protective health behaviors among the general public (Cipolleta et al., 2022).

The news media has played a major role in introducing wastewater surveillance to the public. As early as April 2020, U.S. news agencies were describing the research being conducted to track SARS-CoV-2 in sewers (Barber, 2020; Snyder and Cullinane, 2020). Each new COVID-19 surge

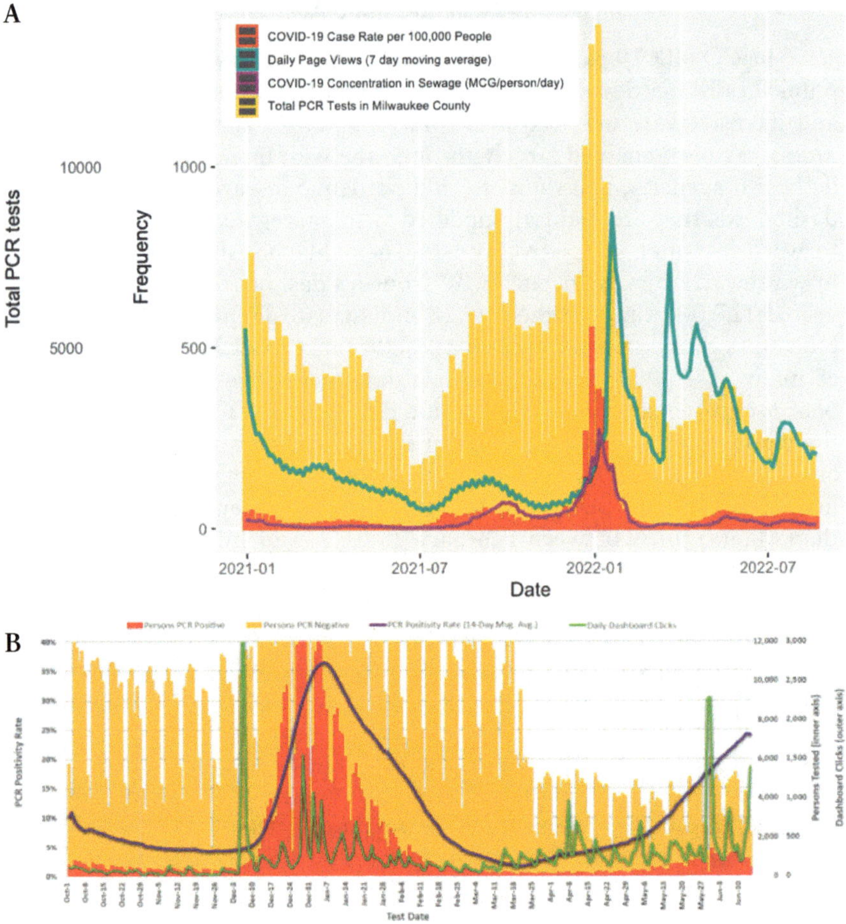


**FIGURE 2-12** Example data on wastewater public dashboard views over time and clinical testing results from (a) Wisconsin's wastewater dashboard and (b) Houston's wastewater dashboard compared to clinical case data. Increased clinical cases and percent positivity is associated with increased page views.
SOURCES: Wisconsin Department of Health Services and City of Houston Health Department.

and variant prompted multiple news stories on the application and value of wastewater surveillance. Although this undoubtedly increased public knowledge on the topic, little is known about how media attention has affected public understanding, perception, and use of wastewater data or how wastewater surveillance has attracted readers compared to other COVID-19 surveillance reporting.

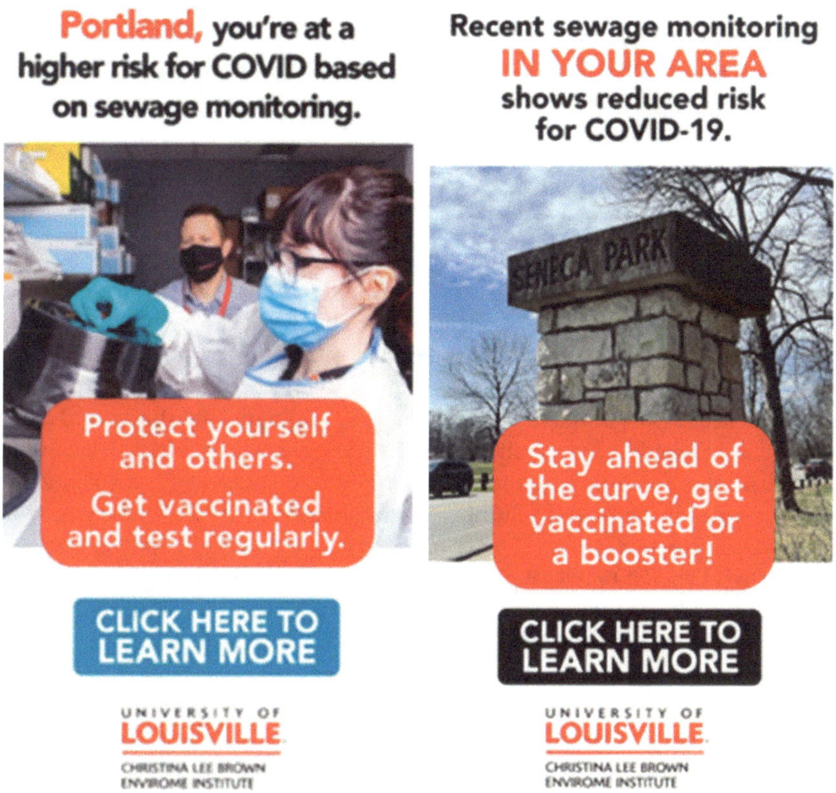


**FIGURE 2-13** After an ad campaign in the highest-risk sewersheds, the number of COVID-19 tests increased compared to other similar-sized sewersheds.
SOURCE: Ted Smith, University of Louisville, personal communication, 2022.

With respect to public awareness and perception of wastewater surveillance for SARS-CoV-2, a study of Louisville, Kentucky, residents in August 2021 found that 43 percent were aware that SARS-CoV-2 could be detected in their sewer system; 49 percent said they did not know whether that was possible (Holm et al., 2022a). Approximately one-third of respondents were familiar with the idea that the amounts of SARS-CoV-2 in wastewater reflected the level of community infections. Less than one-third knew that Louisville's wastewater was being monitored. It is worth noting that the public dashboard had only been available for about 3 months. The same study found that 85 percent of those surveyed were supportive of wastewater surveillance. Those who were aware that SARS-CoV-2 could be measured in wastewater were more likely to be supportive of the activity. In a follow-up survey in early 2022 (LaJoie et al., 2022), the team conducted an online survey for adults across the United States. Similar to the Louisville results, 42 percent of residents were aware that SARS-CoV-2 could be detected in wastewater and 95 percent supported monitoring for diseases in wastewater.

## INNOVATION IN RESPONSE TO IMPLEMENTATION CHALLENGES

The rapid spread of SARS-CoV-2, globally and across the nation, initiated an emergency response to surveillance, including the early implementation of wastewater surveillance. The early voluntary nature of the response initiative, the need for rapid implementation without existing guidance, strained laboratory capacity, and a lack of validated methods and sampling designs were major challenges in developing coordinated surveillance. Innovation, across utilities, analytical laboratories, and health departments, was essential in the rapid scaling and continuous process improvement of SARS-CoV-2 detection.

Early in the COVID-19 pandemic, efforts to collect and analyze wastewater data began with proof-of-concept work that was concentrated in university laboratories familiar with environmental monitoring due to available expertise and laboratory equipment. At the time, there were no straightforward mechanisms to support the validation and scale-up of wastewater surveillance by researchers. As a result, teams took creative approaches to funding their work, including through the National Science Foundation's (NSF's) Rapid Response Research (RAPID) research grants[8] and internal funding provided by the universities and local agencies. Philanthropy also played a role in funding several projects, with sources including the Rockefeller Foundation, CDC Foundation, and others. Over time, more funding sources became available, such as grants from CDC's Epidemiology and Laboratory Capacity for Prevention and Control of Emerging Infectious Diseases (ELC) funds for state, territorial, and eligible local health departments.

Even after CDC funding through the ELC program was launched through health departments, there was little initial guidance for utilities that were interested in participating, and partnerships between utilities and public health departments needed to be developed. The launch of the NWSS in

[8] NSF RAPID grants are relatively small research grants that, according to NSF, are "used for proposals having a severe urgency with regard to availability of, or access to data, facilities or specialized equipment, including quick-response research on natural or anthropogenic disasters and similar unanticipated events." These proposals undergo internal merit review only and so are awarded in a faster time frame than typical NSF proposals (NSF, 2009).

September 2020 created a network of health departments and utilities supported by CDC with peer support for those developing surveillance systems.

Early researchers also dealt with several challenges with respect to laboratory methods and infrastructure, sampling design, and data interpretation. Prior to the pandemic, non-enveloped, enteric viruses dominated the wastewater research world (Polo et al., 2020). An enveloped respiratory virus, SARS-CoV-2 caused early methodological hurdles because many of the wastewater concentration methods had not been validated and optimized for enveloped viruses. Early multilaboratory studies were needed to optimize, validate, and compare these methods (e.g., Pecson et al., 2021). Prior to these studies, identification of appropriate surrogates and controls for PCR-based methods and where to acquire them proved difficult. The need for a central document outlining the minimum quality control information for PCR-based experiments in environmental samples was apparent, and several documents have since been released (e.g., Ahmed et al., 2022; Borchardt et al., 2021).

Laboratory capacity was an issue during the initial phase of the pandemic. Clinical testing was the priority at health laboratories, and most of the environmental microbiology academic and research laboratories were initially closed for the pandemic response. CDC's biosafety recommendations for laboratory requirements to process wastewater samples created further hurdles as many environmental laboratories now had to meet biosafety Level 2 or higher. Supply chain issues also affected almost every aspect of life during the pandemic, including the availability of sampling kits, laboratory reagents, and autosamplers.

Sampling design considerations were complicated early on by the unknown SARS-CoV-2 environmental persistence and infectivity, and uncertainty about whether special precautions for sample collection, transport, and storage were needed. Sample location and frequency decisions were also an early hurdle, and appropriate data analysis and visualization had almost no precedent. To generate actionable data that can be compared across sewersheds, strategies to transform the data to account for dilution effects and compare data spatially have taken time to develop and are not fully resolved. The committee's Phase 2 report will review sampling approaches, analytical methods, data analysis, and data visualization and research needs in light of experience gained during the first 2 years of the pandemic.

## CONCLUSIONS

**The experience with wastewater surveillance during COVID-19, demonstrates that these data are useful for informing public health action and that wastewater surveillance is worthy of further development and**

**continued investment.** Public health agencies that invested in collecting, analyzing, and interpreting the data found them useful to inform policy decisions (e.g., masking and social distancing rules) allocation of public health resources (e.g., testing and vaccination sites, public notification efforts), and allocation of clinical resources (e.g., staffing, hospital beds). These data rarely stood alone but rather were frequently used in conjunction with other disease surveillance data sets (e.g., case and hospitalization counts), each with their own limitations and advantages, to decide on appropriate public health actions and resource allocations. Information on changing SARS-CoV-2 RNA levels in wastewater were shared with the public, often on dashboards, to help inform personal decision making. The launch of CDC's NWSS created an expanded network of utilities and health departments monitoring SARS-CoV-2 in wastewater, thus strengthening COVID-19 surveillance capacity and providing support and guidance for sampling, analyzing, and interpreting data for public health action.

**Wastewater surveillance data have been particularly useful for understanding SARS-CoV-2 trends and the spread of variants.** Wastewater surveillance provides a method to passively assess trends in COVID-19 burden in a community unbiased by the availability of testing or test-seeking behavior. As COVID-19 clinical testing and reported case data became less reliable in recent months due to many factors, including increased at-home testing, wastewater surveillance provided continued information on where the virus was circulating and the degree of exposure risk. Although wastewater surveillance is not currently being used as a standalone method to reliably estimate the number of community infections, SARS-CoV-2 wastewater data have correlated with case data and other conventional public health surveillance data. Depending on a number of factors, including wastewater sampling frequency, the time required for sample transport and analysis, and the time required for data reporting, wastewater SARS-CoV-2 viral trends have the potential to be reported more quickly or along a more consistent time frame than conventional disease surveillance reporting. Wastewater surveillance also provides comprehensive information on the relative proportions of known variants, and genome sequencing of wastewater samples is an effective strategy to screen for emerging variants among a large contributing population, thus providing information in advance of clinical testing data.

**The emergency response to the COVID-19 pandemic spurred innovation and rapid development and implementation of wastewater surveillance; the challenge is now to unify sampling design, analytical methods, and data interpretation to create a truly representative national system while maintaining continued innovation.** Early challenges in initial surveillance sites focused on analytical capacity, sampling design, and data interpretation. The pandemic inspired innovation, which led to a diversity of approaches and methods rather than a single standard method. To date, sites within the NWSS have been based primarily on wastewater utility and public health jurisdiction willingness to participate, including volunteering time and resources, and thus do not comprise a representative national system. Importantly, participating sites have successfully built new partnerships across professional communities with limited prior interactions, spurring innovation and increased efficiency. The challenge is to formalize these roles and partnerships and ensure national representation with best practices for collection, analysis, and use of the data.

# 3

# Vision for National Wastewater Surveillance

This chapter presents a vision for a national wastewater surveillance system that can be a critical asset for early detection of emerging pathogen outbreaks and for monitoring the spread and virulence of existing pathogens. Key elements of a robust national wastewater surveillance system are explained, along with criteria for expanding the pathogens monitored by such a system beyond SARS-CoV-2. Finally, the committee reviews spatial and temporal sampling approaches consistent with this national strategy for surveillance.

## BENEFITS OF SUSTAINED NATIONAL WASTEWATER SURVEILLANCE

Investment in a robust national wastewater disease surveillance system is important to increase national preparedness for emerging infectious diseases and to monitor resurgences of known agents. The key advantage of wastewater surveillance is that it does not rely on clinical testing, thus enabling early detection of disease when clinical testing is not prevalent or when some patients exhibit mild or no symptoms and thus do not undergo clinical testing. Early detection of emerging infectious diseases is critical, as we can control diseases much more effectively when spread in the human population is limited. Thus, early detection can make the difference between the occurrence of a manageable disease outbreak and the progression of a full-blown epidemic or pandemic. The initial investment in wastewater surveillance infrastructure, developed as a result of the COVID-19 pandemic, has already been successful in identifying other emerging threats to public health. For example, wastewater surveillance has enabled early detection of poliovirus outbreaks in New York and London as well as the recent spread of mpox (de Jonge et al., 2022; Nelson, 2022). This critical infrastructure and expertise can provide the foundation of a national wastewater infectious disease surveillance system.

An overriding lesson of the past 2 years is that an outbreak of an emerging pathogen will be followed by a period of remarkable uncertainty. Maintaining a national wastewater surveillance system ensures readiness to respond to evolving risks. Even well into the COVID-19 pandemic, given the unpredictability of SARS-CoV-2 variant emergence and spread, the value of these data to public health management continues to increase. To date, SARS-CoV-2 variants have emerged and spread in global "sweeps" whereby a new, highly transmissible variant that is capable of evading established immunity rapidly spreads. The emergence of new variants of concern is complicated by co-circulation of multiple variants or subvariants concomitantly (Elliot et al., 2022). Variations in human demographics, vaccination rates, and local environmental conditions are highly likely to lead to marked differences in variant/subvariant prevalence across communities (Saad-Roy et al., 2022), with impacts on the need for public messaging and data to inform infection prevention and to enhance hospital preparedness. Representative community-based wastewater surveillance, implemented on a national level with spatial and temporal representation in sampling, can provide data on SARS-CoV-2 trends and variant distribution in a practical and rigorous manner (see Chapter 2).

National wastewater surveillance also has value for monitoring known diseases that vary temporally and spatially. For example, early detection of influenza can provide critical data for healthcare systems and public health messaging in communities. Several illustrative, high-priority use cases that demonstrate the value of a national wastewater disease surveillance system are given in Box 3-1.

**BOX 3-1**
**Potential High-Priority Use Cases for Wastewater Data**

*Detecting a new emerging pathogen and associated variants.* Here, the goal of wastewater surveillance is to monitor for introduction of the pathogen into a new population. For example, the recent report of a case of paralytic poliomyelitis in New York State and the subsequent detection of vaccine-derived polio type 2 in wastewater in several counties suggest community spread, which has not occurred in 40 years (Link-Gelles et al., 2022). Another example is ebolaviruses, which are largely contained to West Africa with only rare cases in the United

## BOX 3-1 Continued

States (WHO, 2015). Surveillance of particular pathogens of interest could be ramped up if there are concerns over importation by travelers when outbreaks have been detected in other countries. In this use case, simply detecting the pathogen may be enough to initiate action (e.g., further testing to determine the extent of spread or to mitigate further transmission).

*Evaluating whether the United States is at risk of an outbreak.* In some situations, one may wish to use wastewater surveillance data to monitor for a new outbreak of a known pathogen potentially already circulating at low or sporadic levels or introduced occasionally via travelers without further spread—for example, an outbreak of hepatitis A or dengue. In these cases, detection of the pathogen alone may not be informative if it is already present, and one may need to evaluate changes in trend or absolute level of the virus instead to determine whether there are signs of increased transmission. This evaluation would be facilitated by a historical record from periodic monitoring that would establish a baseline level and allow comparison with levels during a suspected outbreak.

*Evaluating a variable seasonal pattern.* Characterizing seasonal patterns of disease and timing of epidemic or seasonal onset (e.g., identifying the beginning of flu season) is an important step in ramping up public health resources and messaging (e.g., planning for hospitals' needs or evaluating the timing of annual vaccines). Wastewater surveillance can provide a potentially more sensitive signal for the onset of a regular disease season, if ongoing sampling and analysis is conducted to determine the regular seasonal pattern and compare against current data. Ideally, sampling should be regular enough that thresholds or trend indicators can be used for guidance in interpreting new wastewater surveillance data. For example, if influenza viral load above a certain threshold or a sustained increasing trend for a certain number of days is observed, this may indicate the onset of a seasonal flu outbreak.

*Evaluating transmission levels and trends for an epidemic or a pandemic.* In the case of an ongoing epidemic/pandemic, it may be important to use wastewater data to understand infection/transmission patterns as the nature and utilization of diagnostic tests change over time. For example, in the COVID-19 pandemic, wastewater surveillance has been valuable in providing a more consistent measure of transmission levels when case data become less reliable due to increased reliance on at-home testing and subsequent decreased reporting within the public health system. Wastewater data provide lead time in advance of hospitalizations and/or deaths without requiring active testing of individuals (see Chapter 2). Reliably inferring community disease burden from wastewater data requires calibration with alternative robust sources of data. If sufficiently high correlation is observed, one may be able to use wastewater to infer infection patterns even as the testing rate for cases changes. More broadly, wastewater data can be used to help corroborate trends from multiple data sets with different strengths and weaknesses to understand underlying transmission patterns.

Finally, community-based wastewater surveillance has the potential to provide critical information necessary for understanding the virulence of emergent variants. Virulence—the severity of disease caused by a new variant—can be determined, for example, by dividing the number of hospitalized individuals by the total number of infections in the community. However, as testing for infection has moved from institutionally based testing (the results of which are reported to public health agencies) to at-home testing (in which a significant but unknown number of infections are unreported), a data gap has emerged and the required data for estimating variant virulence at the population scale have been lost. Wastewater surveillance may provide a means to understand trends in disease prevalence relative to trends in hospitalization for indications of virulence, and further scientific advances in using wastewater to estimate the number of infections in a community would enhance this value.

## KEY CHARACTERISTICS OF A NATIONAL WASTEWATER SURVEILLANCE SYSTEM

The committee's vision for a robust surveillance system includes five key characteristics: flexible, equitable, sustainable, integrated, and actionable.

### Flexible

The system should have the flexibility to monitor multiple pathogens at the same time and pivot as needed to new pathogens of public health importance. Both the number of pathogens tracked and the scale of operation (e.g., frequency of sample collection, number of testing sites) should be flexible. For example, an emerging infectious disease threat might require that a system be adapted to test for new pathogens. Similarly, an outbreak of an existing disease in a given area might necessitate an increase in the frequency of testing in that area and expansion of testing to new sites to capture the temporal and spatial attributes of the outbreak.

### Equitable

A robust, useful wastewater surveillance system should be as equitable across population demographics as possible. Equity requires a fair distribution of the benefits and burdens of public health interventions across individuals and communities. Although ethical analyses of public health surveillance systems often focus on their burdens and risks, such interventions can also confer important benefits on individuals and communities (WHO, 2017). For example, wastewater data can form the basis for allocating increased resources and outreach to such communities to improve health (Hrudey et al., 2021; Ram et al., 2022; see Chapter 2). However, if a wastewater surveillance system is not equitable in its coverage, allocating resources using wastewater surveillance data may mean diversion of scarce public health funds away from certain vulnerable populations.

The potential for inequity is a particular concern for wastewater surveillance because there are presently large geographic differences in where surveillance is being conducted (and where it is possible to conduct surveillance) within the United States. Unsewered households (16 percent of the U.S. population; U.S. Census Bureau, 2022) and facilities are by necessity excluded from wastewater surveillance and many, but not all, lie within rural areas. A recent California study found disproportionately low availability of wastewater surveillance in designated "disadvantaged" communities and rural areas of the state (Medina et al., 2022). Moreover, as long as participation in wastewater surveillance remains voluntary on the part of state and local officials, who may decline to implement it for a variety of reasons including political considerations or limited capacity, use of wastewater surveillance could potentially widen existing disparities in how well residents of different states and counties are served by public health systems and programs (Adhikari and Halden, 2022).

An equitable wastewater surveillance system would invest resources in outreach efforts to engage officials from communities that are not currently participating. This engagement should include some assessment of barriers to participation followed by efforts to reduce these barriers when feasible and advantageous. Information about the logistics and advantages of wastewater surveillance for disease detection, as well as a playbook for starting up a local program, could lower barriers to broader participation. Dialogue may also reveal ethical, social, political, or legal worries that could be assuaged by learning from the experiences of existing wastewater surveillance programs.

A robust national wastewater surveillance system should include strategies by which data can be usefully extrapolated through statistical techniques to communities not covered by the system, including unsewered areas (e.g., based on mobility data and laborshed information on commuting and work patterns). Even with additional statistical analyses, wastewater surveillance data may not provide sufficient information about some regions. The use of multiple disease surveillance data sources can also help ensure equity of surveillance efforts with respect to unsewered households and communities, and regional public health agencies should take these data gaps into consideration when investing resources.

### Sustainable

The COVID-19 pandemic highlighted the need for the U.S. public health system to maintain active vigilance in monitoring emerging disease and threats from new pathogens in a way that is sustainable for decades to come. Sustainability of the system will require attention to two issues: financial support and operational burden. Implementing partners, including local utilities, wastewater treatment plants, academic research centers, and public health departments, that have been strained to provide wastewater surveillance services during a time of emergency cannot be expected to do so indefinitely without attention to the burdens that participation involves.

In addition to scaling up financial support, there may be a need to consider ways to scale down operational aspects of wastewater surveillance relative to the pandemic period while maintaining institutional capacity. For example, reduced frequency of sampling may feasibly provide limited baseline surveillance until an emerging threat or disease presence is detected, at which time sampling could be scaled up for more comprehensive coverage. Thus, a sustainable system is one with sufficient support—financial, technical, and logistical—to provide an "everyday" level of sampling and analysis and to scale up for particular pathogens as signals emerge that more intensive sampling or sampling of a different set of sites is needed.

Finally, a sustainable system will require outreach to policy makers and the public to demonstrate the societal value of a wastewater surveillance system to achieving public health outcomes. This, in turn, requires interpretable and actionable data.

### Integrated

There are two key aspects of integration that are essential in a robust wastewater surveillance system: (1) collaboration and coordination across the participating partners (e.g., utilities, analytical laboratories, and public health departments), and (2) the analysis of data from different disease surveillance systems to ensure comprehensive understanding in supporting public health action. Collection of timely and accurate data requires collaboration across and within utilities collecting the samples, laboratories conducting the analysis, and the relevant local and national agencies using and disseminating the results. Data from all three of these system participants need to be fused to generate coherent information. For example, information from utilities and laboratories is needed to determine the accuracy of the data, assess whether there were other substances in the wastewater that could affect the validity of the results (e.g., if industrial chemicals were present that could cause degradation of the pathogen), and provide other information necessary to interpret the findings. Similarly, integration of efforts across different divisions *within* public health agencies (e.g., communicable disease, environmental health, and communications) is critical to support timely and effective action on the part of local, state, tribal, and federal stakeholders using wastewater surveillance system data. As the field of wastewater surveillance is evolving, it is also important to integrate the latest advances and the expanded knowledge base from researchers (Hoar et al., 2022).

Integration with other relevant data is also key for data interpretation to drive public health actions. Wastewater data should not be interpreted in isolation and need to be integrated with information on epidemiological context—both to evaluate trends and hot spots and to understand the risk factors and population characteristics of the communities reflected in wastewater catchment areas. Integrated analysis and interpretation of wastewater data would consider these other important data sets and compare wastewater findings with existing epidemiological surveillance systems (e.g., syndromic surveillance, clinical data). Each of these systems has strengths and weaknesses, but wastewater data can be used in a complementary manner with data across different surveillance systems to understand the underlying population patterns. Seamless integration is particularly important for consistent and effective public health messaging and risk communication, especially when many independent entities are contributing to the collection and interpretation of wastewater data.

## Actionable

The ultimate goal of a surveillance system is to produce actionable data for public health agencies and policy makers. In some cases, data from surveillance systems may even be used to inform decisions that individuals and families make about their risk behaviors (e.g., whether to travel to an area impacted by a disease outbreak). In order to be actionable, wastewater surveillance data must be timely, available, reliable, representative, and interpretable.

*Timely* information is critical for actions to contain infectious diseases. It is much easier to control a disease at the earliest point in an outbreak than when it is already widespread in the community. For early warning potential of wastewater surveillance to be realized, sample collection, analysis, and interpretation of data by public health decision makers must operate on a timescale that allows for informed and timely interventions. Timeliness is related to sustainability in that expeditious collection, analysis, and interpretation of wastewater data are unlikely to occur in the absence of sufficient human and financial resources.

Data are *available* when shared with public health agencies and others who can use the data to support decision making. The question of what availability should look like with respect to other stakeholders, including academic researchers, private companies, and the general public, is more complex. As is discussed in Chapter 4, increased data access requires thoughtful consideration of ethical concerns about privacy and potential complexity or uncertainty in the interpretation of the data. It is clear from the nation's experience during the COVID-19 pandemic, however, that knowledge generation and innovations can be accelerated when data are widely shared.

*Reliability* refers to consistent confidence that the results, with respect to magnitude and trend, represent the viral load and not a product of variability in the system from other factors. If results are not reliable, public health officials run the risk of taking costly actions to combat an emerging threat that does not exist or failing to detect one that does. In addition to financial costs, these outcomes can erode public trust in the public health surveillance system. Therefore, public health officials also need to understand clearly the strength of the evidence and limits on reliability, where they exist. To promote reliability, protocols used to generate the data need to be rigorously validated and reproducible, and the method performance should be openly reported. Reliability can be improved by integrating wastewater data with relevant data (i.e., syndromic data) to improve the sensitivity and specificity of information available and through additional scientific research.

*Representative* wastewater data capture information that is reflective of or is relevant to the population at risk. A key advantage of wastewater surveillance is that it does not rely on individuals seeking diagnostic testing for their symptoms or health conditions. However, to ensure that data from national wastewater surveillance are representative, sampling sites need to be representative of the nation's population.

Finally, wastewater surveillance data should be *interpretable* in a public health context. The analysis methods and interpretation guidance should link the data with population patterns of disease so that public health officials and the public understand what the wastewater data imply for public health. This may necessitate different analytical approaches for different use cases. For example, the simple detection of an emerging pathogen might be actionable if the goal is to monitor for the introduction of a new or emerging disease. On the other hand, for a pathogen that is more established in a community (e.g., SARS-CoV-2, influenza), it would be valuable to develop an analytical approach that links temporal or concentration patterns of wastewater data to patterns of disease inferred from syndromic data (e.g., trends in cases, numbers of cases). Interpretability also naturally relies on integration across utilities, laboratories, and public health, as described above. To support data comparisons and interpretation across geographic areas or at the national scale, the wastewater surveillance should ideally generate data that can be standardized, either at the time of initial collection and analysis or through statistical adjustments.

## A FRAMEWORK FOR IDENTIFYING CANDIDATE PATHOGENS FOR WASTEWATER SURVEILLANCE

Wastewater surveillance holds clear promise beyond SARS-CoV-2 and the COVID-19 pandemic. The question is not whether to sustain and expand wastewater surveillance efforts but how. Selection of candidate pathogens for surveillance necessitates careful consideration.

Application of a defined set of criteria can help guide and systematize evaluation of the most promising candidates for wastewater surveillance. Similar approaches have been proposed for prioritization of candidate pathogens for wastewater surveillance elsewhere (Eaton et al., 2021). Three closely linked key criteria are proposed here for evaluating potential targets for inclusion in a national wastewater surveillance panel (see Figure 3-1):

1. public health significance of the threat,
2. analytical feasibility for wastewater surveillance, and
3. usefulness of community-level wastewater surveillance data to inform public health action.

As discussed in Chapter 2, public health actions could include informing public health resource allocations, informing clinical resource allocations, and/or informing masking, social distancing, and stay-at-home policies. These three criteria are by no means exhaustive, but they form the key pillars of an initial assessment of the value and feasibility of potential expansions of wastewater surveillance. This assessment should be informed by the latest information on existing and emerging pathogens. In the face of limited resources, the criteria can help prioritize among potential candidates and guide efforts toward more in depth consideration of the most promising candidates for surveillance. These criteria can also be used to identify and prioritize research needs for promising candidates that lack key information or analytical methods needed for adoption in the National Wastewater Surveillance System (NWSS).

### Criterion 1: Public Health Significance of the Threat

Pathogens considered for wastewater surveillance would need to pose a significant public health threat (actual or potential) to outweigh the cost and effort associated with the added workload for public utilities and public health jurisdictions. The U.S. Department of Health and Human Services, the National Security Council, and federal departments and agencies have coordinated on the development of national action priorities and strategies for surveillance based on public health significance (ASPR and NSC, 2018). Similarly, the World Health Organization (WHO) maintains lists of priority

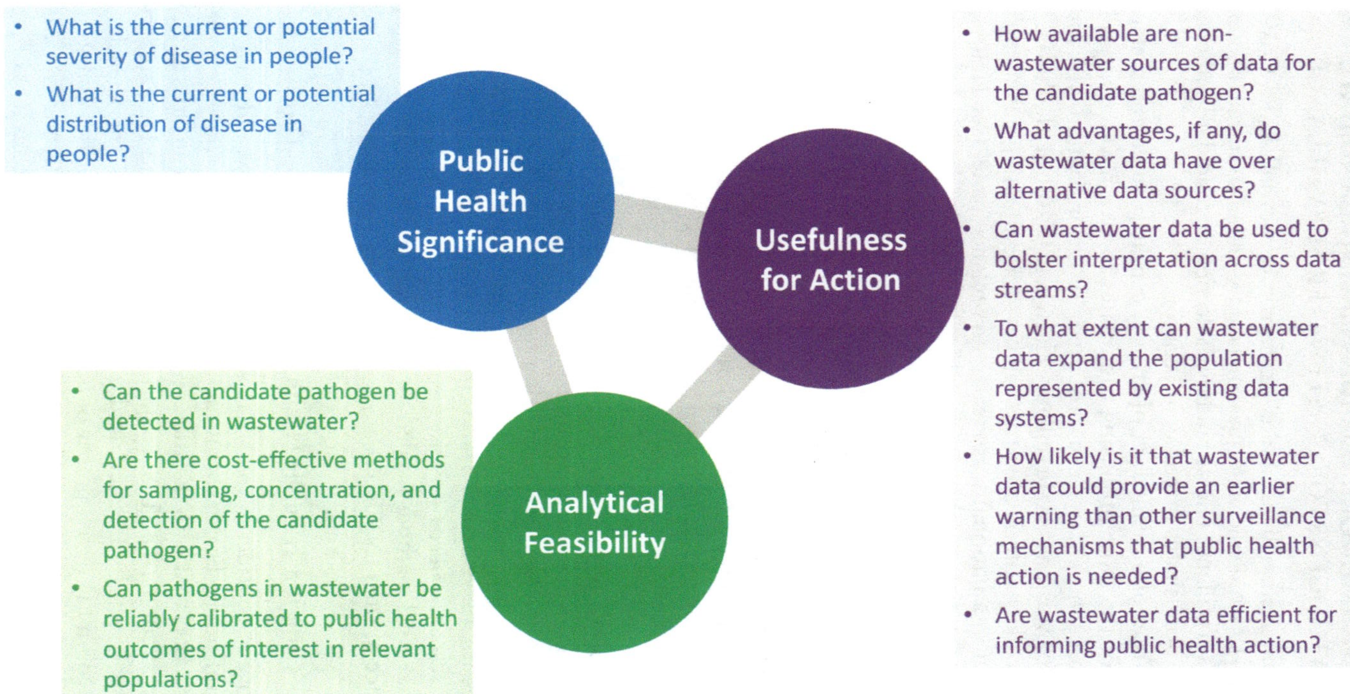


**FIGURE 3-1** Framework for identifying candidate pathogens for wastewater surveillance.

pathogens and diseases, including antibiotic-resistant bacteria, as part of its public health emergency and preparedness strategies for international disease threats. U.S. states also maintain lists of priority conditions (e.g., notifiable conditions).

Drawing on work by these organizations, key parameters to evaluate whether a candidate pathogen for wastewater surveillance meets the criteria for public health significance include the following:

- *What is the current or potential severity of disease in people?* Morbidity, mortality, case fatality, health-adjusted life years lost, or years of livelihood lost are the main parameters characterizing the impact of pathogens in populations along a disease severity spectrum. Disease severity may change over time as new variants emerge, vaccines and treatments are developed or made more widely available, healthcare system capacity increases or decreases, or disease dynamics in the population shift (e.g., the age groups with the highest infection rates). Therefore, this determination should be revisited periodically.
- *What is the current or potential distribution of disease in people?* The extent of disease, as determined by measures of prevalence in the population, is an essential consideration. Surveilling for pathogens that cause rare diseases has limited relevance to public health and presents unique challenges to analytical feasibility (e.g., the quantity of a rare pathogen is unlikely to be detectable at community-level surveillance). The prospect of rapid community spread, with epidemic or pandemic potential, as well as the availability of effective vaccines or therapeutics, should also be weighed for candidates that are not currently widely distributed but have the potential to emerge. Similarly, it is important to consider how much uncertainty exists in our knowledge of the distribution of the disease; for example, diseases that are mostly asymptomatic but cause severe outcomes in a proportion of individuals may have higher uncertainty in the true underlying distribution of disease. Wastewater surveillance may provide an avenue to resolve such uncertainty and understand the extent of spread. Finally, the distribution of disease within the population should be assessed. The case for wastewater surveillance may be especially strong if vulnerable populations are missed or underrepresented in other surveillance approaches. Furthermore, if disease is significantly burdening population groups that have low access to vaccines and healthcare, the argument for heightened investment in surveillance and prevention is strengthened.

### Criterion 2: Analytical Feasibility for Wastewater Surveillance

Wastewater poses unique challenges to detection of pathogens (or pathogen biomarkers; see Box 1-1); therefore, the feasibility and optimization of detection methods needs to be considered fully when evaluating the usefulness of this type of surveillance for potential pathogen candidates. Key questions to assess include the following:

- *Can the candidate pathogen be detected in wastewater?* The pathogen must be excreted in urine/feces or otherwise routinely shed to wastewater for consideration.
- *Are there cost-effective methods for sampling, concentration, and detection of the candidate pathogen?* Sample acquisition and handling should be safe, effective, and readily achievable for public utilities, and detection methods should be precise and reproducible to enable standardization across sites and over time. Evaluation of detection methods should include specificity (i.e., the extent to which the agent is discernible from other targets) and sensitivity (i.e., the potential to detect the agent if present) in wastewater across a range of expected prevalence. Both rare and widespread pathogens pose unique challenges to detection and interpretation of surveillance data. Rare pathogens may fall below the level of detection at the community wastewater scale, while widespread pathogens require quantification rather than simple presence/absence for interpretation. Cost of sampling and testing could vary considerably for different candidate pathogens, with efficiencies achieved as individual specimens are tested for multiple pathogens. The costs for alternate (non-wastewater-based) surveillance methods for the candidate pathogen should be taken under consideration. If detection methods have not yet been optimized for wastewater, the feasibility and cost of method development should be considered.
- *Can pathogens in wastewater be reliably calibrated to public health outcomes of interest in relevant populations?* Although all infectious disease biomarkers will degrade in wastewater, the rate of degradation should be slow enough (or well-characterized enough) to facilitate reliable interpretation of the data. Other contributing sources of the agent in wastewater, such as animals or environmental sources, could reduce the usefulness of wastewater surveillance for public health decision making.

### Criterion 3: Usefulness of Community-Level Wastewater Surveillance Data to Inform Public Health Action

Decisions about the expansion of wastewater surveillance should consider the usefulness of wastewater data relative to other types of surveillance data (e.g., existing, planned, or possible). Key questions in assessing the marginal value of wastewater information include the following assessments. In all of these, the objective is not to replace alternative sources of surveillance but to maximize the proportionate value of wastewater data with these other sources.

- *How available are non-wastewater sources of data for the candidate pathogen?* The extent to which other sources of data are readily available, cost-effective, and capable of informing public health actions should be weighed relative to the potential contributions of wastewater data. For many targets, samples and data could be available at clinics, hospitals, and other healthcare facilities through routine, syndromic-based, or targeted surveillance efforts. For other targets, the samples and data will not be available or it will not be cost-effective to obtain them. Air filter monitoring is increasingly available in public spaces for airborne pathogens and could serve as another source of samples and monitoring data, although air filter monitoring is typically conducted at smaller scale (typically building level) than community-level wastewater surveillance (Bhardwaj et al., 2021; Sousan et al., 2022). Digital disease surveillance (e.g., symptom-based monitoring through online searches, pharmacy data) is another source of information for monitoring disease symptoms (and disease distribution) in populations (Lu et al., 2019; Zhang et al., 2019). For novel or emerging pathogens that cannot be surveilled effectively in other settings, wastewater surveillance could have an especially important role in detection.
- *What advantages, if any, do wastewater data have over alternative data sources?* Even among pathogens for which healthcare data or other sources of information are available, the usefulness of that information may vary. For example, infected persons may be more or less likely to present for care. Wastewater data may have more consistent ascertainment (i.e., the fraction of infections captured in the data) than sources that rely on potentially changing behaviors around care-seeking and testing. If only syndromic data are available, wastewater surveillance data may be able to distinguish between different pathogens that lead to the same symptoms. Wastewater data may also shed light on key pathogen and disease parameters not ascertainable from other data sources. Wastewater

samples can, for instance, distinguish among different variants, strains, and types of biomarkers if analytical techniques enable this level of specificity.

- *Can wastewater data be used to bolster interpretation across data streams?* Wastewater data are particularly useful in comparison with information emerging from other disease surveillance systems to increase confidence in the understanding of trends. Thus, an additional important metric to consider for wastewater surveillance is whether it could provide a functional alternative or a distinct data stream that can be used to validate signals for detection of emerging trends for a specific disease in a timely manner.
- *To what extent can wastewater data expand the population represented by existing disease surveillance systems?* Because wastewater surveillance data can be captured passively, without requiring individuals to seek care, participate in a study, or even exhibit symptoms, they may provide an opportunity to expand the population represented in existing data. Use of wastewater surveillance could be especially pertinent to unique, at-risk, or vulnerable communities that are likely to be missed through other surveillance approaches. Wastewater surveillance can also be particularly useful for diseases where mild or asymptomatic cases are common, or when clinical testing is not widely available or used.
- *How likely is it that wastewater data could provide an earlier warning than other surveillance data that public health action is needed?* This prospect will depend on several factors influencing the ability to monitor real-time trends in the community, including (1) the timing of the onset of pathogen shedding into wastewater; (2) the period of infectiousness; (3) the clinical course of disease (e.g., if individuals shed pathogen into wastewater before onset of symptoms); and (4) other time-varying parameters related to sample processing, detection, and reporting for a given pathogen. Lead time from reporting wastewater surveillance data to observed health outcomes (e.g., positive cases as measured in the same community, clinical disease in local healthcare settings, and local hospitalizations) will be relevant benchmarks for comparing wastewater surveillance data to other data sources for use in informing public health action.
- *Are wastewater data efficient for informing public health action?* It will be important to evaluate whether wastewater data are cost-effective as a replacement, as an additional source of data, or as the only available source of data. Cost-effectiveness should be weighed relative to public health importance and usefulness for decision making as prioritized above.

In some situations, the answers to these questions may be clear (e.g., if no other data about a given pathogen of interest exist). In others, determining whether to invest in wastewater surveillance data for a specific pathogen as part of a national surveillance system may require research studies to evaluate the quality, usefulness, and cost of wastewater data compared to or in tandem with other data sources. Regardless, these criteria and questions can guide how and when wastewater surveillance data may fill a gap or complement other forms of disease monitoring and public health data.

## ILLUSTRATIVE APPLICATIONS OF CRITERIA

Putting the criteria outlined above into action requires careful consideration of the public health significance of the threat, the analytical feasibility of measuring the agent in wastewater, and understanding how this type of information might complement existing public health strategies to monitor the threat and inform decision making. Below, the committee presents several examples of agents of current public health concern and evaluates them against each of the criteria based on the state of the science at the time of writing. Both promising examples and examples of pathogens that do not currently meet the criteria are discussed for illustrative purposes on how the criteria could be applied. This is not intended to replace a thorough evaluation by the U.S. Centers for Disease Control and Prevention (CDC), including assessments of implementation costs relative to the value added beyond other available disease surveillance data. As the state of the science and infectious disease risk evolves, candidate pathogens will need to be re-evaluated; pathogens that may not be good candidates for surveillance now might be well suited for wastewater surveillance in the future.

### Promising Examples

The committee applied the criteria outlined above to several microbial threats that appear to be promising candidates for wastewater surveillance given their public health significance, analytical feasibility of measurement, and value above and beyond existing public health strategies: ongoing COVID-19 surveillance, influenza, antimicrobial resistance, and enterovirus D68 (EV-D68). In some cases, such as for ongoing COVID-19 surveillance, the criteria are mostly already met. In other cases, such as antimicrobial resistance, improvements in features such as analytical feasibility will be important to actualize a useful surveillance scheme.

### *Longer-Term Dynamics of COVID-19 and SARS-CoV-2 Variants*

**Criterion 1: Public health significance of the threat.** SARS-CoV-2 variants continue to emerge and escape established population immunity, posing a significant threat to public health. Although the numbers of hospitalizations and deaths have dropped relative to the number of infections, increased fitness for several of the emerging variants resulted in increased rates of transmission. This resulted in peak hospitalizations during the initial Omicron wave, but hospitalizations in most places in the United States have not risen comparably with subsequent omicron subvariants. Even with hospitalizations and death rates down, the health outcomes associated with SARS-CoV-2 infection in portions of the population have been significant (Ward et al., 2022), and the long-term health burden of "long COVID" continues to be poorly understood. Vaccination and natural infection with prior strains have helped to protect populations, but protection from infection seems to begin to wane after 3–4 months, and protection from severe disease may wane after 6–9 months (Dadras et al., 2022). Uptake of vaccine boosters has not been as consistent as it was for initial vaccination. Another complicating factor is that some newer variants at least partially evade immunity engendered by vaccination or prior infection and recovery. As a result, SARS-CoV-2 continues to be of major public health importance, but if cases of severe disease continue to drop and/or new variants of lower severity and possibly lower rates of long COVID emerge, this criterion will need reassessment.

**Criterion 2: Analytical feasibility for wastewater surveillance.** SARS-CoV-2 wastewater surveillance has been implemented broadly, with demonstrated analytical feasibility of threat detection. As new variants have emerged, they have continued to be detectable in wastewater using reverse transcription-quantitative polymerase chain reaction (RT-qPCR) and RT-droplet digital PCR-based methods that are well established. Furthermore, sequencing-based methods are able to delineate the relative presence of different and even emerging variants in wastewater (Karthikeyan et al., 2022).

A wide range of methods for sample concentration, extraction, and analysis have been demonstrated to be "fit for purpose" for tracking SARS-CoV-2 detection trends in wastewater (Farkas et al., 2022; Maksimovic Carvalho Ferreira et al., 2022; Philo et al., 2022; Wehrendt et al., 2021; Zheng et al., 2022). However, several studies have demonstrated that performance varies considerably between methods (Chik et al., 2021; Pecson et al., 2021); thus, there is now some convergence upon optimal methods. Method standardization would bring greater reliability and validity to the data and also facilitate a broader interpretation of the data across different locations. However, standardization efforts risk stifling further innovation and improvement of methods. An alternative approach to standardization that could provide similar strength without limiting future development is the adoption of performance validation standards (see Chapter 4).

Although detection of SARS-CoV-2 variants is technically feasible, one challenge to expanding this approach for emerging variants is that relatively little information exists on the relative shedding patterns and rates for different variants of SARS-CoV-2 and in the face of partial immunity engendered by prior infection or vaccination. Longitudinal studies in presumably infection- and vaccine-naive individuals infected with the original SARS-CoV-2 strain demonstrated that approximately half of individuals with mild-to-moderate disease shed SARS-CoV-2 ribonucleic acid (RNA) in their feces and that a subset of individuals continue to shed viral RNA for weeks to months after the original infection (Natarajan et al., 2022). Unfortunately, little is known about the dynamics of fecal SARS-CoV-2 shedding in individuals with some level of immunity (through vaccination and/or recovery from prior infection) and with new variants. This information gap confounds strong quantitative assessments. For example, it is possible that as the virus evolves, its shedding may increase, decrease, or completely cease in fecal samples.

Another consideration is that animals may also be infected by SARS-CoV-2; thus, animal waste may also contribute to both human-infecting and non-human-infecting variants that may be detected in wastewater, confounding the interpretation of the results. For example, variant typing in wastewater has identified so-called cryptic lineages containing mutations that have only rarely been detected in human clinical cases (Smyth et al., 2022), which might suggest a non-human contributing source (e.g., rats). A plausible alternative explanation is that these lineages derive from unsampled human infections (e.g., due to persistent shedding by immunocompromised individuals, or infections of different cell types). While these cryptic variants are of scientific interest, it is important to note that these variants currently represent only a minor fraction of the SARS-CoV-2 detected in wastewater. On the whole, analytical feasibility has improved for SARS-CoV-2 in wastewater, with methods available to detect multiple known and emerging variants. However, limitations in understanding shedding dynamics and the presence of non-human viral reservoirs do pose some challenges to interpreting results.

**Criterion 3: Usefulness of community-level wastewater surveillance data to inform public health action.** SARS-CoV-2 wastewater surveillance is increasing in usefulness for informing public health action as SARS-CoV-2 begins to follow more of an endemic transmission pattern, with reduced clinical testing and at-home testing becoming much more common (see Chapter 2). This has led to increased use of wastewater surveillance to understand levels and trends, given its consistent ascertainment. On the other hand, declining rates of hospitalization and death suggest that case (or infection) levels (whether measured clinically or via wastewater) may be less useful to inform action, with hospitalizations and other severity indicators taking a larger role. Nonetheless, given that changes in transmission precede changes in hospitalizations or deaths, wastewater (and case) data remain useful as a leading indicator. Additionally, a major use case for wastewater surveillance of SARS-CoV-2 is variant detection and monitoring, which has been useful for predicting upcoming increases in case load as one variant replaces another as the dominant strain circulating in a population. Wastewater surveillance data continue to be used for public health decision making, staging of resources, and planning and thus are clearly actionable. However, if severity continues to drop as the virus becomes endemic, the actionability of wastewater trend data correlated to case data may diminish. As the infrastructure for wastewater surveillance becomes more mature, the relative cost-effectiveness is expected to improve.

**Summary.** In summary, disease (both acute and long-term) caused by SARS-CoV-2 is still a significant public health concern, particularly for emerging variants. SARS-CoV-2 variant measurement in wastewater is analytically feasible, although detailed information on how shedding may vary in different variants and by immune status (prior infection, vaccination) is, at present, lacking, and non-human reservoirs may also contribute to variants detected in wastewater. SARS-CoV-2 remains a good candidate for ongoing wastewater surveillance, because such surveillance provides information that is likely to be strongly complementary to the more limited clinical testing and variant sequencing that are currently being performed.

### *Influenza*

**Criterion 1: Public health significance of the threat.** Influenza is a prevalent seasonal disease that affects humans and is a significant public health threat. Seasonal drift of the genome of influenza caused by accumulation of point mutations allows influenza to pose an annual public health concern for humans. These annual seasonal outbreaks result in an average of 35,000 deaths and 200,000 hospitalizations in the United States, and between 290,000 and 650,000 deaths globally (Rolfes et al., 2018; Thompson et al., 2004). The 1918–1919 influenza pandemic resulted in 50 to 100 million global deaths. Influenza is unquestionably a significant public health threat. Beyond humans, influenza viruses also infect a number of additional host species with significant potential for enzootic and epizootic transmission. Avian influenza can be characterized as either low pathogenicity avian influenza or high pathogenicity avian influenza (HPAI). HPAI is a devastating agent in commercial bird flocks that can result in the loss of tens of millions of birds and associated finances. Cross-species co-infection with multiple influenza strains can also result in a reassortment of the viral RNA segments, leading to antigenic shift and human pandemic potential for the virus (Kim et al., 2018).

**Criterion 2: Analytical feasibility for wastewater surveillance.** The influenza pathogen can, for the most part, be readily detected in clinical samples using molecular methods such as RT-PCR. These methods have been extended to wastewater, demonstrating that surveillance is analytically feasible. Influenza can be detected in a variety of body fluids. Despite predominant transmission by the respiratory route, influenza is also associated with gastrointestinal symptoms, has been detected in the feces of some infected patients, and thus is likely to be present in wastewater (Wolfe et al., 2022; Ye et al., 2016). Avian influenza can be transmitted by the fecal-oral route in birds (Alexander, 2007). The virus is enveloped and thus somewhat biochemically similar to SARS-CoV-2, so it may also be expected to partition to wastewater solids as does SARS-CoV-2 (Ye et al., 2016). Concentration and recovery methods that have been demonstrated as "fit for purpose" for SARS-CoV-2 should also work for influenza, although this requires performance validation. Detecting and differentiating specific influenza viruses, including those that emerge seasonally via antigenic drift and those that emerge from antigenic shift and have a high pandemic potential, should be achievable using existing molecular approaches and next-generation sequencing.

Influenza virus possesses a single-stranded, negative-sense, segmented RNA genome, and this segmented nature poses a potential complication in strain typing, as multiple strains are likely to be present in wastewater. However, this is no greater problem than delineating different variants of SARS-CoV-2 and may be overcome with next-generation sequencing approaches.

One potentially complicating factor is the wide availability of a live attenuated influenza vaccine (e.g., Flumist). This vaccine replicates in the body and is likely excreted to wastewater through feces or respiratory secretions. The potential for live vaccine to confound wastewater surveillance would need to be evaluated.

Although additional research is warranted to better understand the frequency, level, and pattern of shedding to wastewater; the persistence of the virus in wastewater; and the performance of wastewater methods for broader community surveillance, influenza outbreaks have already been monitored by wastewater surveillance on a local or institutional scale (Mercier et al., 2022; Wolfe et al., 2022). As a result, there is strong evidence for the analytical feasibility of detecting influenza in wastewater.

**Criterion 3: Usefulness of community-level wastewater surveillance data to inform public health action.** Influenza infections are prevalent, and because only a small subset of infected individuals present for clinical care and are captured by public health surveillance systems, wastewater surveillance data are expected to be of high use in informing public health action. Furthermore, dominant strains that infect humans often change on a seasonal basis. Although some clinical typing of strains is done, this information may lag behind initial upticks in case rates. Thus, new strains of influenza may first be detected through wastewater surveillance instead of clinical typing; furthermore, once a new strain of influenza is found to be circulating in a population, wastewater surveillance could supplement clinical and syndromic surveillance to provide early warning of spread of the virus to new regions. Even tracking of seasonal flu within populations using wastewater surveillance has the potential to inform public health decisions on communication to the public and distribution and staging of resources (i.e., vaccine clinics, hospital staffing).

**Summary.** Influenza is a good candidate for potentially expanded wastewater surveillance. Flu-like diseases caused by influenza are a significant public health concern because certain strains cause significant disease, especially in immunocompromised populations, and influenza has previously caused pandemics. Influenza measurement in wastewater is analytically feasible, and wastewater detection has already been demonstrated to be useful in limited studies, although use of a live-attenuated vaccine may complicate interpretation of influenza measurements in wastewater. Wastewater surveillance may provide a complementary source of information to existing methods for influenza strain and case rate tracking using clinical information that enables better understanding of overall case rate as it includes individuals with less severe or no symptoms. Furthermore, wastewater data might serve as a leading indicator of the emergence of new influenza strains in some settings.

### *Antimicrobial Resistance*

**Criterion 1: Public health significance of the threat.** Antimicrobial resistance is a critical threat in medicine and has been declared as 1 of the top 10 health threats facing humanity by the WHO (EClinicalMedicine, 2021).[1] CDC's *Antibiotic Resistant Threats in the United States, 2019* estimates that there are 2.8 million resistant infections annually in the United States, responsible for 35,900 deaths (CDC, 2019). In addition, *Clostridioides difficile* infection, attributable to antibiotic disruption of the gut microbiome, is estimated

[1] See https://www.who.int/news-room/fact-sheets/detail/antimicrobial-resistance.

to cause an additional 12,800 deaths each year (CDC, 2019). Newly emergent threats such as drug-resistant *Aspergillus* and *Candida auris* infections highlight the continually evolving nature of the threat. The direct healthcare costs associated with six common multidrug-resistant bacterial pathogens were determined to be $4.6 billion per year, with *C. difficile* representing another $1 billion (CDC, 2019; Nelson et al., 2021). Globally, an estimated 4.95 million deaths were associated with antibiotic-resistant bacterial infections in 2019 (AMC, 2022). Thus, antimicrobial resistance is of great public health significance (see also NASEM, 2022).

**Criterion 2: Analytical feasibility for wastewater surveillance.** Detection of antimicrobial resistance in wastewater is technically feasible, although some challenges need to be overcome to make surveillance robust, extensible, and reliable. Detection of antimicrobial resistance is routinely performed in clinical bacterial and fungal isolates using culture and sensitivity-testing approaches. Additionally, molecular-based (PCR) methods are also used to amplify and detect mutations in genes that confer antibiotic resistance as well as mobile genetic elements that can confer antimicrobial resistance. Many of the molecular methods that are used to detect antimicrobial resistance may be adaptable for wastewater surveillance. Genes or transcripts encoding antimicrobial resistance mechanisms are commonly present at detectable levels in wastewater, including those of CDC-listed "urgent" and "serious" threats. The gut microbiome has been demonstrated to be an important reservoir of pathogens that infect other tissues in humans (e.g., bacteria can translocate from the gut to the blood and cause bloodstream infections; see Tamburini et al., 2018). Thus, shedding of pathogens and their antimicrobial resistance genes would be expected to occur often in individuals harboring these pathogens. The wastewater resistome (defined as the collection of antimicrobial resistance genes and mutations detected in wastewater) would be predicted to be relatively stable over time (Brinch et al., 2020), suggesting that infrequent sampling and testing could be used to assess shifts in the prevalence of antimicrobial resistance over time. Detection of antimicrobial resistance at the point of wastewater treatment provides a community-wide analysis of current prevalence and trends.

The wastewater resistome will reflect the predominant resistance genes in the community microbiome, predominantly the human gastrointestinal microbiome—although human non-intestinal microbiomes and animal microbiomes may also contribute to detectable levels depending on the community. One challenge to using wastewater surveillance for the presence, absence, and abundance of antimicrobial resistance genes is that it reflects genes that are present in both pathogens and commensal organisms (which are carried without causing disease in most individuals). Thus, the wastewater resistome as detected by amplification of specific resistance genes or transcripts will largely reflect commensals circulating in a predominantly healthy population. Linking resistance genes to specific pathogens is certainly feasible but, in most cases, will require either long-read sequencing or bacterial culture on selective media followed by resistance gene amplification—more costly, labor-intensive approaches. Thus, neither approach is amenable to routine wastewater surveillance but may be useful if triggered by specific hospital or community disease outbreaks (see Box 3-2).

Many clinically relevant antimicrobial resistance genes are carried on plasmids, which are pieces of DNA that replicate separately from the bacterial chromosome and may be transferred between related and sometimes unrelated organisms. Detection of these antimicrobial resistance elements may be relevant whether they are found in commensal organisms or in pathogens. This is because the overwhelming majority of multiantibiotic resistance is carried on mobile plasmids that can move from non-pathogenic commensal bacteria to pathogenic bacteria. Additionally, even commensal

**BOX 3-2**
**The Unknown Source of a Colistin-resistant *E. coli* Infection**

Antibiotic-resistant genes circulate throughout human and animal populations as well as in the shared environments but often stealthily until serious infections are detected within a healthcare setting. The emergence of the first described colistin-resistant infection in the United States illustrates this challenge. Colistin is frequently considered a "last-resort" antibiotic in the treatment of multidrug-resistant Gram-negative bacterial infections in humans. Although banned for animal use in numerous countries, including the United States and Canada, colistin has been widely used as a feed additive antibiotic in livestock in low- and middle-income countries. In 2015, the detection of *mcr-1*, a gene encoding colistin resistance on a mobile plasmid, was first reported in China (Liu et al., 2016). In 2016, an *mcr-1* mediated colistin-resistant *E. coli* infection was detected in a single hospitalized patient in Pennsylvania (Kline et al., 2016). The patient had not traveled outside of the United States in the prior year and had no exposure to livestock, raising the question of whether this *mcr-1* mediated resistance had been circulating stealthily in the community prior to hospital-based detection. CDC and the Pennsylvania Department of Health screened 105 individuals with known contact with the patient, including in the community and in healthcare facilities—none were identified as carrying the *mcr-1* gene. Wastewater surveillance, not available at the time, would have been a valuable adjunct to directed contact tracing in determining whether colistin-resistant bacteria were circulating in the community. Testing archived or contemporary wastewater samples from the community and hospital point sources would provide both broader population coverage and cost-effectiveness relative to directed contact tracing.

bacteria can cause severe infections in certain immunocompromised individuals when carried into a hospital or acute care nursing facility. CDC has prioritized carbapenem-resistant *Acinetobacter* and carbapenem-resistant *Enterobacterales* as "urgent threats" and extended spectrum beta-lactam-resistant *Enterobacterales* as a "serious threat" (CDC, 2019). The genetic determinants responsible for encoding these resistances are carried on mobile plasmids and are commonly detected within commensal bacteria within communities—thus amenable to wastewater surveillance and actionable in terms of informing healthcare facilities of the risk in susceptible patients. This value is supported by research that has shown that wastewater detection roughly parallels detection of resistance in hospitals within the wastewater catchment (Parnanen et al., 2019).

**Criterion 3: Usefulness of community-level wastewater surveillance data to inform public health action.** Wastewater surveillance data on antimicrobial resistance are likely to be useful to inform public health action, though the exact ways in which they complement existing data sources and result in specific actions have yet to be defined. Thus, while promising, wastewater surveillance of antimicrobial resistance may not yet be ready to put into use at this time. Although the wastewater resistome is expected to be relatively stable over time, there are specific use cases when targeted sampling and analysis might be highly valuable. An example is the detection of a previously unknown multidrug-resistant infection in a clinical setting (see Box 3-2). Relatedly, as new resistant bacteria have frequently been detected outside the United States (e.g., NDM-1 beta-lactamase, which confers broad resistance to beta-lactam antibiotics; Yong et al., 2009), targeted screening at community wastewater facilities linked to or serving international points of entry would provide an early awareness signal to healthcare facilities and clinical laboratories. In U.S. clinical laboratories, detection of resistance is largely automated and screens for currently known resistance patterns based on "antibiograms," which use local or regional aggregate data. The emergence of a new resistant mechanism would not be included in automated testing and requires use of a non-automated test. Awareness of emergence detected in wastewater would alert clinical laboratories to screen for resistant bacteria otherwise undetectable based on historical antibiogram data. Analysis of human waste collected from incoming aircraft has provided geographically defined patterns of resistant determinants and may represent a source of introduction into a community (Nordahl Petersen et al., 2015). These data would complement data from the Global Sewage Surveillance program that has been testing sewage for antimicrobial resistance markers in 60 countries since 2016 (Aarestrup and Woolhouse, 2020; Hendriksen et al., 2019).

**Summary.** Fecal shedding of pathogens and their antimicrobial resistance genes is expected to occur, and their measurement in wastewater is analytically feasible. However, the high prevalence and range of antimicrobial resistance genes present in commensal organisms may make it difficult to identify relevant increases of antimicrobial resistance genes above the very high background rate that is detected. Antimicrobial resistance is of high public health significance, and if analytical advances allow more rapid and cost-effective mapping of resistance to specific pathogens, surveillance in wastewater would be of even higher value. Furthermore, as antimicrobial resistance frequently emerges in "hot spots" outside the United States and then spreads globally, wastewater surveillance at sentinel sites such as airports may serve as an early warning signal for additional screening of newly emergent antimicrobial-resistant pathogens. In summary, antimicrobial resistance is a promising candidate for future wastewater surveillance system development, though some challenges and exact applications have to be further investigated and defined.

### *Enterovirus D68*

**Criterion 1: Public health significance of the threat.** EV-D68 is a non-polio enterovirus that is of moderate public health significance. EV-D68 has circulated in the United States since at least 1962 and has been linked to biennial seasonal outbreaks in the fall since 2014 (Helfferich et al., 2019). The virus is associated with a range of clinical illness, ranging from mild acute respiratory disease to a severe polio-like paralysis (acute flaccid myelitis, AFM) (Sooksawasdi Na Ayudhya et al., 2021). Although the virus is a relatively rare cause of significant respiratory disease, the unprecedented severe cases in young children have made this a virus of significant public health concern, despite the lack of clear information on its prevalence, distribution, and transmissibility. This virus is spread through respiratory droplets via direct person-to-person contact and indirect contact through touching of contaminated surfaces, raising concern for high transmissibility. The virus is expected to demonstrate a high degree of asymptomatic infection.

**Criterion 2: Analytical feasibility for wastewater surveillance.** Detection of EV-D68 in wastewater is expected to be analytically feasible, though the few existing methods for its detection in wastewater are not quite as developed as they are for other enteroviruses such as poliovirus. Enteroviruses commonly infect humans, and existing clinical methods that leverage RT-PCR for virus detection are used to detect viral strains of particular concern in patients. Thus, it should be relatively straightforward and feasible to expand these existing approaches to develop a wastewater surveillance approach for the virus. In fact, a recent study by Tedcastle et al. (2022) described detection of EV-D68 in wastewater samples in the United Kingdom; furthermore, studies from Israel and Scotland have already demonstrated wastewater surveillance for EV-D68 (Erster et al., 2022; Majumdar et al., 2019; Weil et al., 2017).

In addition to RT-PCR approaches for virus detection, sequencing of the viral envelope protein 1 (VP1) region of the viral genome allows EV-D68 to be distinguished from other enteroviruses. Thus, it is feasible to develop a next-generation sequencing approach to both detect and characterize the virus in wastewater. Such an approach would require amplification of a specific region of the genome (amplicon) and subsequent sequencing of that region to both detect and differentiate this virus from other closely related enteroviruses. As a non-enveloped RNA virus that can likely be detected using standard RNA extraction and either sequencing or RT-qPCR or droplet digital RT-PCR methods, EV-D68 would be easily adaptable to cost-effective inclusion in a wastewater surveillance panel.

Like other enteroviruses, EV-D68 infects the gut and thus may be detected in stool. However, the level and magnitude of shedding is not clear. Although little is reported on the stability or persistence of EV-D68 or components of the virus such as RNA or antigens in wastewater, as a non-enveloped virus, EV-D68 and its RNA would be expected to be relatively stable. Past studies on other enteroviruses have shown that they may be stable for days to months in wastewater. At least one prior study in Israel used both clinical and wastewater-based disease surveillance in the investigation of EV-D68 (Erster et al., 2022). Furthermore, the biennial pattern of outbreaks since 2014 suggests an ability to design a study to calibrate the environmental signal to clinical outcomes. No animal reservoirs of EV-D68 have been described; thus, no other contributing sources to wastewater are expected other than human infections.

**Criterion 3: Usefulness of community-level wastewater surveillance data to inform public health action.** Wastewater surveillance data about EV-D68 are expected to be highly complementary to the limited existing public health data on this pathogen and are expected to inform public health action. Because EV-D68 infection often is asymptomatic or minimally symptomatic and severe respiratory disease or AFM as a consequence of EV-D68 is not reportable from a public health perspective, little is known about the prevalence and seasonality of this virus in the community. Thus, inclusion in wastewater surveillance could contribute significantly to understanding the biennial pattern of outbreaks and raise awareness for the virus as an etiologic agent in the face of other more commonly circulating viruses. Identification of an uptick in EV-D68 could, for example, help to warn hospitals to watch for clinical manifestations of the virus. It also could help to stage resources for epidemiological investigation.

**Summary.** The diseases caused by EV-D68 are a significant public health concern. EV-D68 measurement in wastewater is analytically feasible, and shedding occurs at a high enough rate that detection has already been demonstrated to have value in limited studies. Given the lack of alternative methods to measure EV-D68 prevalence and burden in different communities, wastewater surveillance is expected to be of high possible usefulness. In summary, EV-D68 is a promising candidate for expanded and improved wastewater surveillance.

### Examples of Pathogens That Are Not Currently Applicable or Need More Data

Applying the criteria outlined above, many microbial threats do not currently meet the level of public health significance, analytical feasibility of measurement, and adequate value above and beyond existing public health strategies required to be considered for broad implementation. As an example of how the criteria above might be applied to evaluate such candidates, two examples—*Candida auris* and prions—are explored below in detail. In these examples, it is evident that some criteria are fully or partially met, but others fall substantially short. Both of these currently questionable candidates are evaluated below to illustrate how to apply these three criteria in public health decision making as it relates to candidate selection.

#### *Candida auris*

**Criterion 1: Public health significance of the threat.** *Candida auris* is an emerging fungal pathogen (yeast) that can cause a range of infections from mild superficial infections to severe invasive infections. *C. auris* is of high public health significance as it has been linked with serious outbreaks in healthcare settings and is frequently resistant to commonly used antifungal drugs. *C. auris* is a significant risk for those in clinical settings and nursing homes, particularly those undergoing invasive procedures, but infections have been detected across age groups. Due to the multidrug resistance, this is a significant pathogen of concern. However, most cases have been linked to hospital- or care facility-based exposures, and little is known about colonization of individuals outside the clinical setting. *C. auris* seems to be highly transmissible in a clinical setting as evidenced by reported outbreaks. It is expected to be spread by person-to-person contact in healthcare settings but has been shown to be persistent in the environment and may also be indirectly spread by contaminated surfaces in healthcare settings. The disease may be severe, with death resulting in as many as one in three cases.[2] *C. auris* now has international distribution, having been reported in half of the U.S. states and more than 30 countries.

**Criterion 2: Analytical feasibility for wastewater surveillance.** It is unclear whether existing analytical methods and the level of shedding of *C. auris* into wastewater are adequate for detecting this microbial threat in wastewater. Detection of *C. auris* in the clinic is performed by fungal culture and mass spectrometry, or by molecular PCR-based methods that amplify and then sequence the D1–D2 region of the 28S ribosomal RNA gene (rRNA) or the internal transcribed spacer (ITS) region of the rRNA. Although culture-based methods are likely not expandable to wastewater surveillance, amplicon sequencing-based methods may be valuable for detecting *C. auris*, if it is present at a high enough concentration to be detected. Recent research has demonstrated that *Candida* species can be detected in hospital wastewater (Mataraci-Kara et al., 2020), although whether such organisms can be detected in community wastewater treatment facilities remains unknown. Thus, although some likely portable molecular methods exist for detection, considerable research is necessary to demonstrate the feasibility of methods, and the presence and persistence of the yeast in wastewater. Currently no methods have been described for the isolation or detection of *C. auris* from wastewater. Direct extraction of the fungal nucleic acid from wastewater may be feasible, but it is not clear whether *C. auris* would be present at high enough levels to allow detection.

The presence of *C. auris* in human stool is not well described in humans, although mouse studies suggest that fungal burden in stool is higher for some invasive strains of *C. auris* than for other strains (Abe et al., 2020). Because the fungus can colonize the skin of individuals, *C. auris* shed with sloughed skin (e.g., from hand washing or showering) may be a source of loading into wastewater. Based on the available data describing known niches for *C. auris* infection and colonization in humans, no other contributing sources of *C. auris* to wastewater have been demonstrated. The best available data also suggest that the concentrations of *C. auris* in wastewater are likely rather low, which might challenge detection, even with amplification-based methods. Furthermore, it is unclear whether the agent is stable or perhaps can replicate in wastewater, though it has been demonstrated to persist on environmental surfaces.

Finally, because colonization can occur in the absence of infection, it is unclear whether wastewater data would be correlated with clinical outcomes, and no data exist to suggest that levels in wastewater relative to clinical outcomes would be consistent over time. That being said, as

[2] See https://www.cdc.gov/fungal/candida-auris/candida-auris-qanda.html.

colonization rates increase, wastewater data may correlate with increased exposure for susceptible hosts and thus increased infections.

**Criterion 3: Usefulness of community-level wastewater surveillance data to inform public health action.** Information on *C. auris* presence and abundance in different geographic communities is expected to be of high usefulness, as little other information is available regarding the distribution and abundance of this organism. *C. auris* became a notifiable condition in 2018 in the United States. At present, limited institutional screening for *C. auris* is being carried out, mostly in individuals who are strongly suspected of being colonized with the fungus. As infection with the agent can range from mild to severe, it is expected that the current rates of *C. auris* infection that are reported are an underestimate of the actual population-based burden of infection or colonization with the pathogen. Thus, wastewater surveillance data, if accurate and quantitative, might be helpful. Furthermore, very little is known about community prevalence of the agent outside clinical settings. It is possible that wastewater surveillance could help inform a better understanding of the distribution of the agent, but it is still not clear how it would be actionable. For the time being, most cases of *C. auris* seem to occur in hospital and long-term acute care settings; thus, more proximal and local wastewater surveillance at these institutions may be preferable to broad-scale regional wastewater surveillance.

**Summary.** *C. auris* is a questionable candidate for wastewater surveillance at present, given limitations in analytical feasibility of detection and the fact that it is unclear how much this organism is shed into wastewater. The infections caused by *C. auris* are a significant public health concern, and the broad-scale antifungal resistance of this pathogen makes it an agent of particular concern. Additionally, there is a lack of alternative, thorough methods to measure *C. auris* prevalence and burden outside of hospital and long-term acute care facilities where patient screening might be performed. If research and development resolve the uncertainties and challenges for *C. auris* detection in wastewater in low-prevalence settings, wastewater surveillance might be of moderate to high value, particularly if the population prevalence of *C. auris* increases in coming years.

### *Prions*

**Criterion 1: Public health significance of the threat.** Prion diseases are rare but can be very severe, and as such, they are of moderate overall public health significance in most settings. Prions are transmissible proteins with an abnormal conformation. They can trigger an abnormal folding of native cellular proteins in the brain, resulting in transmissible spongiform encephalopathy diseases (TSEs). These diseases primarily target the central nervous tissue and are usually rapidly progressive and fatal. Both human and animal TSEs exist, with concerns about the potential for cross-species transmission. In animals, the spread of the proteins is believed to be through bodily fluids, either via direct contact or indirectly through a contaminated environment. In humans, most TSEs are genetically inherited or arise spontaneously within an individual, but there are types of prion disease that result from infections acquired from others or the environment. Specifically, transmission of the rare but severe Kuru disease and variant Creutzfeldt-Jakob disease is believed to be through ingestion of infected meat. Due to the rare nature of the human disease and the poor transmissibility of the agent, most TSEs are unlikely to pose a broad and significant public health concern. However, chronic wasting disease in deer may pose a risk to hunters and others handling or consuming felled deer, and many states in the northern Midwest have surveillance for chronic wasting disease using neural tissue from killed deer.

**Criterion 2: Analytical feasibility for wastewater surveillance.** There are substantial challenges to the analytical feasibility and robust and reliable detection of this agent in wastewater. Diagnosis of prion disease is made by immunocytochemical or protein-based detection methods performed on brain biopsies. PCR-based methods are not applicable because the agent is protein based. Prion-protein detection can be carried out by standard biochemistry methods such as immunoblotting (Nicholson, 2015; Yokoyama, 1999). To date, no methods have been described for the detection of prions in wastewater. Detection of prion proteins in wastewater might be possible through antigen-based detection (e.g., enzyme-linked immunosorbent assay, ELISA) in matrices that have prion protein present. Alternatively, there is a protein misfolding cyclic amplification assay (Green and Zanusso, 2018), but it is not clear how this would be feasible to adapt to wastewater. Furthermore, it is unclear how the prions would be isolated from wastewater samples for analysis.

Although there are reports of fecal shedding of prions by deer and goats (Cheng et al., 2016; Haley et al., 2011; Krüger et al., 2009; Miller and Williams, 2004; Safar et al., 2008; Tennant et al., 2020; Terry et al., 2011), there are no reports of fecal or other shedding of prions in humans (although it is plausible). It is not clear that animal prions could be differentiated from human prions in wastewater, though presumably methods such as ELISA can use antibodies that differentiate between the different proteins that cause animal versus human prion diseases. Should differentiation of prions from animal versus human sources be challenging, it is important to note that animal wastes are unlikely to be a large contributing source in most wastewater systems, although prions from rodents, domestic pets, and wastewater flow from slaughterhouses could be contributing sources. Prions are extremely stable in the environment; they resist degradation, disinfection, and treatment processes. If present in sewage, they would be expected to persist.

**Criterion 3: Usefulness of community-level wastewater surveillance data to inform public health action.** At present, wastewater surveillance data on prions are not expected to significantly contribute to informing public health action. Most TSEs are not transmitted but rather are inherited or arise spontaneously within an individual. Due to the fact that TSEs are rapidly progressing conditions, infected individuals would likely seek out medical care, and as a notifiable condition, information would be conveyed to state and national public health agencies. It is also unclear how any wastewater surveillance data could be broadly acted upon by public health agencies due to the rarity of the disease.

**Summary.** The diseases caused by prions are severe and important but of low current prevalence with poor transmissibility and therefore of lower general public health concern. Prion measurement in wastewater is not currently analytically feasible, although biochemical methods exist that could enable method development, and the data are currently expected to be of limited value for public health management of prion diseases.

## VISION FOR AN EFFECTIVE FRAMEWORK FOR DETERMINING TEMPORAL AND SPATIAL RESOLUTION

When developing a vision for a national wastewater surveillance program, a key consideration is its spatial and temporal resolution—that is, how frequently and from which locations samples should be collected to provide optimal cost-effectiveness and usefulness of the data. The temporal and spatial resolution of sampling needs to be determined based on the objectives of the program (e.g., the use cases in Box 3-1), which will be a function of the pathogen(s) of interest, the analytical methods, and both the epidemiology and pathogenesis of the disease. A typical set of objectives for the overall surveillance program could be determination of infection and disease prevalence and ascertainment of "hot spots" and/or "hot moments." The relative variability of a target pathogen in space and time and the value of understanding this variability needs to be balanced with the expense of collecting and analyzing at different sites and times. Temporal and spatial variability will be discussed, followed by a potential path forward to design an overall sampling strategy.

### Temporal Variability

Temporal variation is an important consideration in determining a sampling frequency that is affected both by changes in flow within a sewer system as well as the timescales involved in changes in disease transmission patterns. For example, different pathogens have different speeds at which an outbreak might progress or seasonal patterns that affect the needed frequency of sampling over the course of the year. Each of these sources of temporal variation will affect the sampling frequency needed to capture useful wastewater surveillance information for different pathogens.

It is well known that influent flow rate and composition coming into wastewater treatment plants fluctuate owing to daily and weekly variations in contributions to the sewer system, which can affect pathogen measurements in wastewater (Wade et al., 2022). Additionally, the relative extent of these variations tends to be larger in smaller treatment plants owing to an averaging effect that occurs in the collection system (Tchobanoglous et al., 2003). In wastewater collection from combined sewers or older leaky sewers, there could also be variability with storm events.

As an example, Li et al. (2021) took daily samples of liquid and solids from raw wastewater following the imposition of more stringent public health measures in August 2020 (see Figure 3-2). There was clearly a downward trend, corresponding to attenuation of community spread of COVID-19, but there were also daily fluctuations about the trend of about 1–2 logs. Mendoza Grijalva et al. (2022) conducted hourly sampling for SARS-CoV-2 in the influent to a treatment plant in Contra Costa, California. They observed a greater likelihood of detection during periods of peak diurnal flow (under dry weather conditions) and recommended collaboration with the wastewater utility to ascertain timing of usual peaks. The fluctuations suggest the reasonableness of daily composites, preferably flow weighted, to achieve detection, although this finding needs to be checked for any new pathogens that are studied. The frequency of sampling would need to be determined based on the use case and the pathogen of concern.

Infectious disease transmission also changes at a range of different timescales, depending on the characteristics of the population and the pathogen (Anderson and May, 1992; Delamater et al., 2019). Population characteristics such as the level of population movement, population density, and age and other demographic factors that influence susceptibility and immunity may all affect transmission patterns, causing outbreaks and other changes in transmission patterns to occur over different timescales. For example, an outbreak could proceed rapidly through a highly susceptible, dense population but move much more slowly through a population with lower contact rates and higher immunity. Additionally, each of these

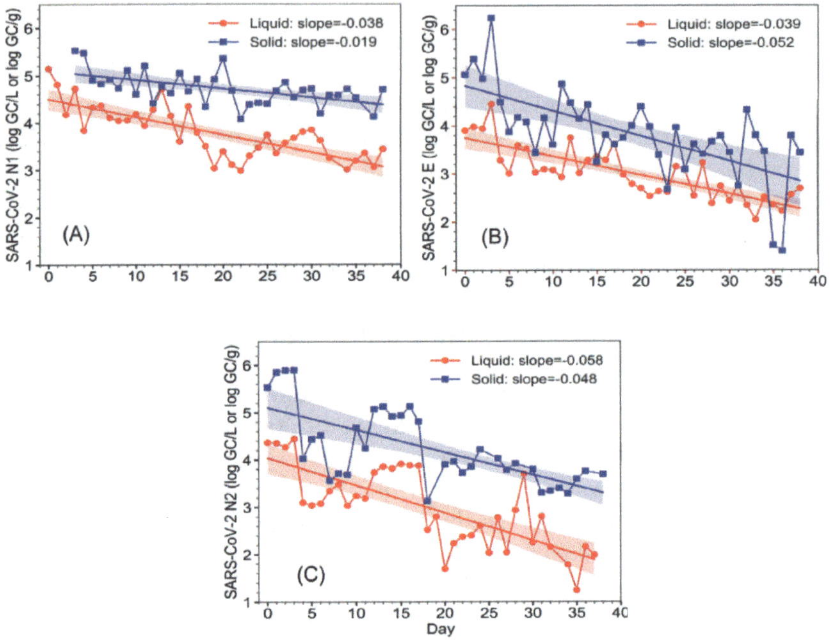


**FIGURE 3-2** SARS-CoV-2 RNA concentration dynamics in wastewater from the Sand Island wastewater treatment plant (Honolulu, Hawaii) following August 2020 "lockdown." Liquid fractions shown in red and solid fractions in blue with data reported for three gene assays: (A) N1, (B) E, and (C) N2. The shadings represent 95 percent confidence intervals of the linear regressions, and the concentration data have not been normalized. Different rates of decline were observed for the three different gene assays.
SOURCE: Li et al. (2021).

factors may change over time, adding additional complexity to the choice of sampling frequency. Furthermore, pathogens have different transmission routes, incubation periods, infectious periods, and other natural history characteristics that mean that the timescale of changes in transmission for different pathogens can vary widely, with some having outbreaks that span days or a week or two, and others that move through a population over the course of months. Different pathogens have different temporal distributions in when and how long they are shed, and how that relates to the clinical manifestation of disease (e.g., some pathogens are shed before onset of symptoms, some may continue to be shed after symptoms resolve). Further, these temporal shedding patterns may evolve over time as the pathogen evolves.

All of these factors will impact how rapidly an outbreak will spread and how frequently samples should be taken to determine important changes in disease transmission. For example, changes that occur over months may only need weekly sampling, while a rapidly evolving outbreak where action is needed more immediately might require semiweekly or daily sampling. The useful sampling frequency for a given pathogen may vary as a function of the season or upon detection (e.g., less frequent sampling and analysis that shifts to frequent sampling if a pathogen is found).

### Spatial Variability

Spatial factors can result in variability in wastewater surveillance data over regional and national scales. For example, samples from wastewater treatment plants serving larger populations do not provide the same level of granularity as those from a plant serving smaller populations (Sharara et al., 2021), although they do capture a wider population and may avoid some of the noise and variability of plants serving smaller populations. Even if the wastewater treatment plants sampled served the same population size, multiple demographic and environmental factors are spatially heterogeneous, including

- annual precipitation;
- proportion of domestic versus nondomestic flow;
- proportion of flow attributable to hospital, healthcare, and congregate living facilities;
- combined versus separate sewers;
- population demographic and socioeconomic factors (e.g., socioeconomic status, social vulnerability, and urban/rural patterns);
- proportion of people who commute to work or school from outside the sewershed; and
- spatial heterogeneity in disease spread patterns.

These factors may vary at the sewershed or sub-sewershed level (see Box 1-3) or may vary at a range of spatial scales from census tract to regional or state level (e.g., regional laborshed and commuting patterns, or spatial variability in contact patterns that influence disease transmission). This variability in the different epidemiological and wastewater features across space can affect the optimal design of a wastewater surveillance system, and the choice of a spatial scale and resolution needed to accurately capture trends and patterns may vary by pathogen and use case (see example use cases in Box 3-1). For example, if a disease tends toward highly localized, distinct outbreaks or clusters (highly spatially heterogeneous), and population contact patterns mean that there is relatively little spatial mixing of the population, one may need a highly spatially resolved wastewater surveillance system to localize transmission hot spots in a way that is actionable for public health. On the other hand, for SARS-CoV-2 and other pathogens that tend to show broad-scale community spread or where population contact/mobility is high, it may be sufficient to have fewer, potentially larger sewersheds that are monitored and provide information about overall population trends in an actionable way (although analysis of CDC's National Wastewater Surveillance System [NWSS] data would be needed to determine the optimal sampling plan).

Data from the current sampling program and related efforts that measure different pathogens across multiple sites (e.g., mpox, influenza, respiratory syncytial virus [RSV]) could be analyzed to ascertain how spatial variability in the factors listed above may impact the understanding of disease transmission for different use cases (e.g., to discern trends, detect new outbreaks). This information would assist in the design of a cost-effective, representative sampling framework for the nation and help assess the value of an adaptive framework, wherein some combination of spatial scales is regularly monitored but then more finely resolved as needed once a pathogen or trend of interest is detected.

### Considerations for Designing a Representative Sampling Framework

A foundational question in the design of a representative strategy is the articulation of the objectives and their desired weighting. As stated by Cheng et al. (2020),

> Disease surveillance systems are established and designed for diverse purposes, including to collect data for understanding variations in disease frequency across populations, space, and time, to monitor pathogen composition over time, to detect outbreaks and forecast epidemics, to assess the impact of interventions, and to determine risk factors associated with diseases. Most surveillance systems operate with multiple public health objectives. Hence, surveillance system designs should generally be subject to multi-objective optimization, and tradeoffs between different objectives must be considered.

The allocation of wastewater surveillance sampling effort over time and space can also be viewed as an optimization question. Two other questions then arise in the design of a sampling program that has the objective of discerning prevalence and trends:

1. How "dense" in space should the sampling sites be, and how many should be set up?
2. How frequently should each site be sampled?

The temporality/frequency of sampling will depend on the dynamics of the particular pathogen and disease and the objective of the surveillance system. As shown in the Honolulu COVID-19 data in Figure 3-2, the decline in wastewater concentration with time during a "lockdown" was discernible with daily sampling and likely would have been discernable had the sampling been twice or thrice a week. The question for another pathogen would be whether a more rapid discernment of trend has value in terms of public health actionability that would justify the increased effort.

The density of sampling in space depends upon how readily more sparse allocation of efforts can discern prevalence and trends compared to a denser allocation. This may differ in parts of the United States due to demographic differences over geographic areas. In some cases, community mobility may be sufficient so that populations who reside in unsewered areas or in small separate sewersheds may contribute to a large city's wastewater system where they work or attend school, such that centralized wastewater surveillance could still sample this population to some degree. Additionally, statistical tools may be useful to extrapolate available data to unsampled areas. As shown by Fairchild et al. (2013), in the case of clinical influenza sampling in Iowa, it is possible for sparser (and hence less expensive) networks to provide data of similar information value to denser networks. The NWSS and similar data could be analyzed to shed light on these factors.

Consideration of equity of sampling with attention to environmental justice populations needs to be embedded in site selection. Equitable action to improve public health requires information from a disease surveillance system that represents and resolves patterns for a range of populations across sociodemographic groups, particularly those who may not be as well-reflected in traditional clinical surveillance methods that may be impacted by issues of access to and availability of care, issues of trust, and issues of cost. There is the potential for wastewater surveillance to provide a more fully representative lens through which to conduct public health surveillance, as it does not require active participation by individuals. Such data could be used independently or compared with clinical data to understand disease epidemiology. Importantly, there is also an ethical trade-off relating to the spatial scale at which wastewater surveillance is conducted: sampling at the larger community level helps avoid stigmatization of particular neighborhoods and minimize threats to individuals' privacy, but examining smaller sewersheds could help target resources and useful public health efforts to the areas that need them most (see Chapter 4 for additional discussions of ethics and privacy).

A full national picture of disease distribution will require comparative analysis of wastewater data with other sources of information. As discussed previously in this chapter, there are significant populations that are not connected to municipal wastewater systems, and these unsewered populations tend to be more rural. Rural areas also tend to have lower median incomes and reduced access to health services than sewered populations [Indiana ACIR, 2019; Long et al., 2018]. These findings underscore the importance of finding alternative approaches for disease surveillance in these communities to avoid potential health disparities. Furthermore, not all sewered jurisdictions would institute a wastewater surveillance system, due to cost, staffing, or cost-effectiveness considerations or unwillingness to participate. Therefore, community-based wastewater surveillance should be viewed as only one tool in national disease surveillance. An aggregate disease surveillance program can achieve equity, even if any single component, such as wastewater surveillance, may not cover all components. Design of the best complementary tools for the unsewered population is important but beyond the scope of this study.

The bottom line is that the temporal and spatial resolution of a sampling program should be subject to intentional design, should be informed by preliminary and ongoing data, and may differ with different pathogens and use cases. Without thorough statistical analysis of existing data, CDC cannot be sure that, moving forward, its investments are appropriate to achieve representative information.

## Sentinel Sites

Wastewater-based surveillance for SARS-CoV-2 was initially set up and continues to be carried out in a range of settings, from highly localized efforts at universities or prisons to aggregated and more standardized surveillance spanning hundreds of wastewater treatment plants across the nation. Although this report focuses on the importance of a national wastewater surveillance system, hyperlocal surveillance at selected sentinel sites in certain circumstances would be an integral element of a true national system. A sentinel site here refers to a location where enhanced or specific surveillance should take place because it represents the "front line" of entry to a larger community. Wastewater infectious disease surveillance information gathered as close as possible to where targets of interest enter a community may have a differentiated value from that obtained from a non-selective community wastewater surveillance system. For example, wastewater surveillance at major U.S. airports and ports of entry would identify initial cases for pathogens from other regions in travelers entering the country (Agrawal et al., 2022; Medema et al., 2020). At these sites, an abundance of global travelers in a localized setting could serve to enrich targets of interest that may be too dilute to detect in larger wastewater surveillance systems prior to significant disease transmission. Furthermore, it may not be cost-effective to perform surveillance on a broad range of global targets of interest at a national level across a broad network, but focused surveillance of a much larger number of targets could be carried out at these more limited sentinel sites, while serving the interests of the nation as a whole.

Given the unique function of sentinel sites within a larger public health system, these sites may differ from regular network sites within a national system both in terms of the types of pathogens targeted and the frequencies at which sampling would occur. Sentinel sites could serve to screen for a wide range of potential pathogens or diseases of concern in other countries, including emerging pathogens. Initial pathogen detection at one or more sentinel sites could trigger expanded wastewater surveillance at communities or broad-scale national surveillance, as appropriate for the pathogen detected. Because of the need to be responsive to the emergence of potential threats in a timely manner, the frequency of sampling at these sites would benefit from being more frequent and/or dynamic in response to global public health information. Sentinel sites could thus be activated with rapid upscaling of sampling efforts under specific circumstances that would justify enhanced screening.

The process of selecting sentinel sites depends largely on the target that is under consideration. Large airports with a high flux of global travelers may be appropriate sentinel sites for global surveillance of pathogens and antimicrobial-resistance genes that are prevalent outside of the United States. Temporarily ramped up sentinel wastewater surveillance should also be considered to coincide with large-scale gatherings that could facilitate disease introductions or transmission, such as at international sporting events. However, these are not the only types of sentinel sites that should be considered. Because emerging pathogens and antimicrobial-resistance genes can arise from animals, another type of sentinel site that would be important to evaluate is animal-intensive areas such as livestock and poultry farms and large zoos (if runoff from animal enclosures is collected into the sewershed). Because of the need for a variety of sentinel sites based on these targets, it would be advisable to include among these sentinel sites both major cities and small, rural wastewater treatment plants. As diseases emerge, or after they have emerged, correlating local wastewater surveillance data with case and hospitalization data would inform how these types of surveillance systems should scale.

In the committee's vision for a national community-based wastewater surveillance system, the development of sentinel sites has clear additive value. These sites may provide early warning of emerging threats in the United States before they reach the general population and could directly inform subsequent scaling of surveillance for new or re-emerging pathogens in the broader national surveillance system.

## CONCLUSIONS AND RECOMMENDATIONS

**Wastewater surveillance is and will continue to be a valuable component of the nation's strategy to manage infectious disease outbreaks, including continued surveillance of SARS-CoV-2 variants, resurgences of known pathogens, and newly emergent pathogens.** The emergency establishment of wastewater surveillance has proven its value, and the efforts at local and national scales to establish the NWSS provide a solid basis for expanded applications. Infectious diseases, whether endemic, seasonal, newly emergent, or re-emergent, are dynamic and never fully predictable. The high likelihood that SARS-CoV-2 variants will continue to emerge and circulate is alone a strong rationale to maintain and strengthen a national wastewater surveillance system. The recent use of wastewater surveillance for poliovirus and mpox in mid-2022 illustrates the advantages of a maintained national system for detecting re-emerging pathogens and pathogens recently introduced into the United States.

**To achieve its goals, a national wastewater surveillance system should be flexible, equitable, integrated, actionable, and sustainable.** Flexibility includes the ability to track multiple pathogens simultaneously and pivot quickly to new threats. A national wastewater surveillance system should be as equitable as possible across population demographics, with efforts to engage underrepresented communities and extrapolate findings, where feasible, to unsewered communities. Integration, including coordination and collaboration across multiple partners (e.g., utilities, laboratories, and public health agencies) and analysis of data from different disease surveillance systems, ensures effective data interpretation in support of public health decision making. For the information to be actionable and inform decisions about clinical and public health resource allocations as well as policy decisions, it must also be timely, available, reliable, representative, and interpretable. Finally, the system needs to be fiscally and operationally sustainable. Although the NWSS supports both local and national public health decision making, a sustainable national wastewater surveillance program may not serve every locality's objectives but should allow for locally funded initiatives, such as pilot surveillance of a pathogen of emerging regional concern.

**When evaluating potential targets for future wastewater surveillance, CDC should consider three criteria: (1) public health significance of the threat, (2) analytical feasibility for wastewater surveillance, and (3) usefulness of community-level wastewater surveillance data to inform public health action.** Applying these criteria to known and emergent/re-emergent pathogens of concern can guide strategic allocation of effort and resources. Assessment of the public health significance of a microbial threat is important to develop and maintain a system that is responsive to current public health needs. Assessment of the feasibility to detect a specific pathogen in wastewater for disease surveillance is necessary to determine technical readiness and can also drive research or technology development for microbial threats that meet the other criteria. Finally, it is critical that the value of wastewater surveillance information for a given pathogen be considered in the context of the broader universe of surveillance approaches so as to maximize the use of resources to inform public health action (e.g., allocation of clinical or public health resources). Candidate pathogens will need to be re-evaluated periodically as scientific knowledge, technology, and infectious disease risks evolve.

**Temporal and spatial resolution of the NWSS sampling program should be subject to intentional design, informed by rigorous and iterative analysis of data for prioritized pathogens.** Collaborative and frequent analysis of incoming NWSS data is essential to determine the spatial and temporal scales of sampling and analysis needed, both for effective COVID-19 monitoring as well as detection of emerging pathogens. Temporal and spatial resolution should be regularly re-evaluated to ensure the system is capable of detecting meaningful change with sufficient lead time needed to inform public health action. CDC should also give careful attention to the need for more representative sampling for prioritized use cases. Currently, the system consists of localities, tribes, and states that were willing and able to participate during a pandemic emergency, and this current distribution of sampling sites might not be representative of the range of demographic and geographic characteristics desired in a national network nor equitable, optimally actionable, or sustainable. Because 16 percent of the U.S. population resides in unsewered communities, wastewater surveillance in and of itself cannot be fully representative of the population but should be viewed as one key component of a national infectious disease surveillance system.

**CDC should take additional steps to bring the benefits of wastewater surveillance to critical areas not addressed by the NWSS.** The committee identified three steps that CDC could take to ensure that resources expended on wastewater surveillance systems are not distributed inequitably. First, CDC should create a comprehensive outreach program to provide information to selected public health officials and utility personnel in localities that are not currently using wastewater surveillance about the potential benefits of joining the national system. Second, CDC should reduce financial and staff capacity barriers to joining the system. CDC could reduce barriers by providing continued and expanded funding to state, tribal, local, and territorial health departments and utilities and by creating an easily operable data management and analysis system wherein local wastewater surveillance programs can easily transmit their samples and data for centralized analysis and data visualization (see Chapter 4). Finally, because some areas that are important to understanding national infectious disease transmission will remain outside the wastewater surveillance system even with these resources in place (e.g., in unsewered areas), CDC should assess whether tools can be used to extrapolate data from monitored regions to estimate disease burden in areas without wastewater surveillance. CDC and local health departments should also maintain robust infectious disease surveillance programs using other sources of data on disease trends and provide public education about how to interpret wastewater data alongside other indicators.

**As part of a national wastewater surveillance system, strategic incorporation of sentinel sites is recommended as a mechanism for early detection.** Sentinel sites should be intentionally selected to monitor for specific emerging pathogens at their points of entry into human communities. Sites that can directly inform community wastewater-based surveillance, especially as related to emerging pathogens, will provide important and distinct benefits in the context of a national surveillance network. Such sentinel sites could include wastewater surveillance at major international airports with a large number of global travelers to detect emerging pathogens and antimicrobial resistance genes. Sentinel monitoring at ports of entry could allow early detection of emerging pathogens entering the country that otherwise may be too dilute to detect at the community scale. Wastewater treatment plants with zoos or major livestock farms that contribute to a sewer system could also serve as valuable sentinel sites to detect the emergence and transmission of zoonotic pathogens. Developing useful sentinel sites will require careful planning and thoughtful experimentation with site selection, program design, and data interpretation based on the pathogen(s) of interest. Sentinel sites are a cornerstone of any public health system, and the NWSS should seek to incorporate these sites in a way that will ensure the surveillance system is nimble and adaptive as needed to address emerging threats.

# 4

# Strategies for Achieving the Vision and Increasing the Public Health Impact of National Wastewater Surveillance

This chapter addresses how the vision of a robust wastewater surveillance system described in Chapter 3 can be achieved. First, the committee describes a formal process that could be implemented to systematically and transparently evaluate potential targets for wastewater surveillance. Second, legal and ethical issues arising from wastewater surveillance are discussed. Although these concerns are not weighty enough to militate against the expansion of wastewater surveillance at the community level, the committee discusses the benefits of creating an ethics governance structure to consider the potential impacts of expanding the scope of data collection or data sharing. Third, the committee examines the steps needed to assure that wastewater data are reliable, representative, comparable, interpretable, and, therefore, actionable. Fourth, key elements of a sustainable system are examined, including operational capacity and financial dimensions, both of which may require substantial investments. Finally, the committee considers the collaboration and coordination necessary to support an integrated system and the respective roles of federal, state, and local participants in such a system.

## A SYSTEMATIC AND DYNAMIC PROCESS FOR EVALUATING TARGETS FOR WASTEWATER SURVEILLANCE

The U.S. Centers for Disease Control and Prevention (CDC) has the important task of prioritizing among candidate pathogens or pathogenic markers to include in the National Wastewater Surveillance System (NWSS). The previous chapter described three substantive criteria for distinguishing among potential candidates: public health significance of the threat, analytical feasibility, and usefulness of community-level wastewater surveillance data in informing public health action. In this section, the committee proposes a strategy to apply those criteria, including a formal process for generating an initial list of candidate pathogens, prioritizing among them for inclusion in the NWSS, and regularly updating both.

Development of (and updates to) a list of candidate pathogens should build on existing work, evaluating extant lists of candidates against the substantive criteria described in Chapter 3. Several groups already have strong infrastructures for identifying candidate pathogens, including but not limited to CDC; the National Academies of Sciences, Engineering, and Medicine's Forum on Microbial Threats;[1] local and state health agencies that issue reports about new infectious agents; and the World Health Organization's Global Antimicrobial Resistance and Use Surveillance System (GLASS).[2] Additionally, public health significance may be assessed by looking at clinical case and hospitalization data, treatment costs, availability and uptake of vaccines, and lists of antimicrobial resistance genes reported in clinical cases. Community wastewater surveillance targets may also be identified from concerns identified in other countries or pathogen outbreaks in highly localized settings, like hospitals.

CDC should develop and implement an open and transparent process by which potential targets for wastewater surveillance are evaluated according to key selection criteria. A useful analogue is the development of a "contaminant candidate list" (CCL) by the U.S. Environmental Protection Agency (EPA) for drinking water monitoring and regulation. The CCL is a list of contaminants that are known or anticipated to occur in the nation's public drinking water systems but are not currently regulated under the Safe Drinking Water Act of 1974 (NRC, 2001). The CCL is mandatorily updated every 5 years and is used to prioritize contaminants for regulatory decisions. Although there are clear differences between the CCL and a list of agents that could be selected for wastewater surveillance, two aspects of the CCL development process have high salience for making determinations about agents to include in wastewater surveillance: the use of expressly defined criteria and procedures for obtaining public input. Notably, the substantive criteria applied by EPA in making regulatory determinations on contaminants from the CCL resemble the criteria described in Chapter 3 for wastewater surveillance: (1) the contaminant may have adverse human health effects, (2) "there is a substantial likelihood that the contaminant will occur in public water systems with a frequency and at levels of public

---

[1] See https://www.nationalacademies.org/our-work/forum-on-microbial-threats.

[2] See https://www.who.int/initiatives/glass.

health concern," and (3) regulation would provide a meaningful opportunity to reduce health risks.[3]

Based on the recommendations of the National Drinking Water Advisory Council (NDWAC, 2004) and the National Research Council (NRC, 2001), EPA began soliciting public nominations for unregulated contaminants for inclusion beginning with the third CCL. Similar to the development process for the CCL, CDC could solicit public comment on potential candidates for wastewater surveillance, making targeted outreach to the global academic community and experts from public health, industry, and utilities. To update the candidate list, this solicitation could be repeated at regular intervals (e.g., biannually, or more often during periods of disease outbreaks of public health significance). Once a list of potential targets is defined, the public could be further included through posting of requests for information on the agents and/or open public comment period on candidate targets.

From the potential candidate list, CDC will need to prioritize an initial suite of wastewater surveillance targets. In so doing, CDC would benefit from creating a systematic process for drawing on external expertise. For example, CDC could create and regularly convene an external advisory panel of academic, government, and industrial partners to provide input and advice on the proposed list of targets relative to agreed-upon selection criteria. Public communication about CDC's decision making and the independent scientific input to that process is important to help members of the public understand what information is (and is not) being collected about their community and why. In some cases, the decision may be that a pathogen is not yet ready for inclusion because of incomplete understanding of its analytical feasibility in wastewater. In such cases, if the pathogen has current or potential public health significance, CDC should re-evaluate the candidate in the future.

The universe of organisms that can be monitored will evolve over time and with additional knowledge. For example, advances in the application of RNA and DNA sequencing methods and analytical approaches (e.g., metagenomic sequencing) for wastewater samples may also hasten the detection of novel or emerging pathogens. Therefore, a process should be developed to revise the candidate pathogen list as well as the prioritized suite of wastewater surveillance targets regularly, while identifying research and development activities that would help fill important knowledge gaps. This revision process should follow the same practices for inclusion and transparency as the initial selection process. On occasion, global health threats may arise that may necessitate rapid consideration of new wastewater surveillance candidates "off cycle" from the periodic review. This is

[3] See https://www.epa.gov/ccl/basic-information-ccl-and-regulatory-determination.

especially likely if the global threat arises from a novel pathogen. In these instances, the advisory panel could be convened on an emergency basis to provide input and comment to CDC on additional proposed wastewater surveillance targets, with in-depth evaluation at the next regularly scheduled review. Thus, the committee envisions a process that is methodical with opportunities for public and expert input while also having the flexibility to move quickly when the need arises.

## PUBLIC ACCEPTANCE: LEGAL AND ETHICAL CONSIDERATIONS

The success of a national wastewater surveillance program relies on building public trust in the system (Hrudey et al., 2021). This is a difficult time to be expanding public health surveillance because of the politicization of the COVID-19 pandemic and increase in the proportion of Americans expressing distrust of public health agencies. Such a context can make a seemingly benign public health intervention suspicious to many people.

There is also substantial potential for confusion on the part of the public about the purpose of wastewater surveillance, because "surveillance" is a charged term for some communities. Discussions about expanding wastewater surveillance programs are occurring at a time of pervasive and increasing surveillance of the public by both state and nonstate actors, generating heightened public concern about "surveillance creep" (WHO, 2017). Furthermore, the historical context for these discussions matters, and it bears remembering that one of the earliest proposals for wastewater surveillance was related to monitoring of illegal drug use (Sims et al., 2021). The public's sense of what wastewater surveillance involves and what its purpose is may be quite different from the community-level infectious-disease monitoring done in the NWSS.

Finally, there may be confusion and difficulty communicating about the kinds of data that are and are not captured in wastewater—especially human DNA—and the potential for identifying specific individuals and households. All of these considerations elevate the need for both careful analysis and skilled communication of ethical issues. Presently, an incomplete understanding of the legal and ethical implications of wastewater surveillance may be impeding some communities from pursuing it, to the detriment of their residents (McClary-Gutierrez et al., 2021).

In this section, the committee reviews legal and ethical considerations and steps that could be taken to address them. Consistent with the study's focus, this section addresses *community-level* wastewater surveillance of infectious disease. Collection and analysis at a smaller scale, or use of the data for other purposes (e.g., law enforcement), raises additional concerns but is outside the scope of this analysis.

### Legal Issues in Community-Level Wastewater Surveillance

Overall, legal concerns raised by community-level wastewater surveillance are minimal. When done purely for public health surveillance purposes, this data collection and analysis does not constitute human subjects research and thus does not implicate federal and state regulations relating to the design and institutional review board review of such research. Another potential legal issue is whether data may be used by law enforcement officials to identify persons who may be violating the law. However, such use is quite improbable because community-level wastewater surveillance examines samples containing information from thousands of households and buildings. There are more cost-effective ways for law enforcers to build a case against those who violate the law (Hall et al., 2012).

Although some legal scholars have discussed the possibility that courts might consider wastewater surveillance an unconstitutional search or seizure (Gable et al., 2020), such a holding is unlikely. The Fourth Amendment to the U.S. Constitution protects individuals against warrantless searches conducted without their permission but only where the individual has a "reasonable expectation of privacy" in the thing or premises being searched. Courts recognize that people do have a reasonable expectation of privacy in their bodily waste in some situations (e.g., a urine sample that is subjected to a drug test), but several lower courts have held that that reasonable expectation terminates when the waste irretrievably flows into a public sewer.[4] Courts analogize sewage to garbage: once left out on the curb, one has no right to expect it will remain private.[5] Notably, the prior cases on wastewater all involved sampling in a location that would enable the analyst to trace contaminants back to a specific source (e.g., a pipe running from a factory to a public sewer). Because community-level wastewater surveillance involves far lower prospects of identifying particular contributors, the courts' reasoning applies even more strongly to that context. Some scholars have argued that if individual identification through wastewater surveillance becomes possible, a recent decision by the U.S. Supreme Court holding that people can have a reasonable expectation of privacy in their cell phone location data could change courts' analyses of wastewater (Ram et al., 2022). But identifying individuals from community-level wastewater monitoring is not technically feasible at this time.

Even if courts did recognize a privacy interest in wastewater, they would still be unlikely to find that wastewater surveillance for infectious disease violates the Fourth Amendment. Provided that the wastewater data

---

[4] Riverdale Mills Corp. v. Pimpare, 392 F.3d 55 (1st Cir. 2004); United States v. Spain, 515 F.Supp.2d 860 (N.D. Ill. 2007); United States v. Hajduk, 396 F.Supp.2d 1216 (D. Colo. 2005); People v. Elec. Plating Co., 683 N.E.2d 465 (Ill. App. Ct. 1997).

[5] California v. Greenwood, 486 U.S. 35 (1988).

are used only for public health surveillance and not law enforcement, courts would probably hold that the search is legal under the "special needs" doctrine. This doctrine holds that a warrantless, unconsented search can be constitutionally permissible if it is reasonable and there exists a "special need" for the information apart from law enforcement, such as communicable disease control (Gable et al., 2020; Joh, 2021). For these reasons, there is little basis for concern that community-level wastewater surveillance programs pose constitutional problems.

## Ethical Concerns Arising from Community-Level Wastewater Surveillance

Although legal concerns are minimal, two ethical issues arising from community-level wastewater surveillance merit close analysis: ensuring privacy and appropriate use of data.

### *Privacy*

Considering whether wastewater surveillance unduly intrudes on people's privacy is ethically important because the ordinary moral and legal presumption in the United States is that individuals are entitled to control access to information about their health (Gable et al., 2020). Wastewater contains information about many aspects of the health status of a group of individuals, from diseases to substance use to genetic information. When it is possible to link health information to a particular person or household, the concepts of privacy and "dignitary harm" become salient, particularly if the person or persons did not give permission for their health information to be accessed. There may also be risks—social, reputational, economic, or even physical—to individuals or communities that arise from private health information being accessed. For obvious reasons, it is impossible to obtain individuals' informed consent to access their information in a community-level wastewater sample, or to allow individuals to opt out of the sample. It is ethically acceptable to conduct public health surveillance under these circumstances (WHO, 2017), but due attention must be paid to minimizing risks and burdens and assuring population benefit (Klingler et al., 2017).

The extent to which wastewater surveillance raises privacy concerns depends on the likelihood that the health information in wastewater can be individually identified. This, in turn, depends on (among other things) the number of individuals represented in a wastewater sample. The smaller the number, the greater the risk of reidentification. At the present time, when wastewater surveillance is done at the community level (Scassa et al., 2022), the risk of identifying individuals or households that made particular contributions to a sample is ordinarily very low. A typical sewershed contains tens of thousands of households or more, and the personal information collected is both anonymous and aggregated. Therefore, wastewater surveillance can presently be conducted for most communities without troubling implications for privacy.

Sampling on a smaller scale, potentially such as in some sentinel sites or very small sewersheds, involves higher potential risk that particular persons, households, or farms might be identified (CWN, 2020; Gable et al., 2020; Ram et al., 2022), thus meriting additional consideration of ethical and confidentiality concerns and additional data use controls (see below). Sentinel sites themselves should be engaged early in the planning process to ensure that ethical and confidentiality concerns, and other site- or event-specific challenges, are addressed prior to surveillance activities.

Although the degree of privacy intrusion is currently low, public health authorities conducting community-level wastewater surveillance have an ethical obligation to monitor the extent to which the capacity to identify individuals and households from otherwise anonymous data sets ("reidentification") strengthens over time. It is reasonable to assume that at some point in the future, human reference databases will become robust enough to be able to identify particular genes that are more frequent for particular demographic groups, as well as genes that characterize particular individuals. For these reasons, assessment of the risk of identification should be dynamic rather than a "one-and-done" evaluation. Although it may be tempting to adopt simple heuristics (e.g., the notion that individuals cannot be identified if the sample captures wastewater from 10,000 people or more [Sims et al., 2021]), such assumptions require periodic validation.

In addition to advances in reidentification capabilities, several other future developments would merit reconsideration of current assumptions about privacy-related risks. These include (1) a court ruling that wastewater data can be subpoenaed or are admissible in criminal or civil proceedings unrelated to public health interventions—for example, drug-related prosecutions; (2) any report of wastewater data sharing with law enforcement agencies, or use of wastewater surveillance infrastructure by law enforcement; (3) any shifts in the scale of wastewater data collection or analysis from the community level to more targeted surveillance, such as individual farms or particular neighborhoods (Scassa et al., 2022); and (4) changes in the degree of stigmatization or other adverse consequences likely to flow from a finding of high levels of a pathogen in a particular community.

### *Appropriate Use of Data*

Because analysis of community-level wastewater data for disease surveillance raises few serious ethical concerns but other forms and uses of wastewater data are more problematic, ensuring that data are not misappropriated is perhaps the paramount ethical imperative for public health officials. This obligation encompasses several dimensions.

At a basic level, data should be securely stored, with reasonable protections against unauthorized access (hacking). The need for appropriate data security protections is especially great where data from many jurisdictions are consolidated in one platform, creating a single target for hacking or exploitation. Guidance from the Canadian Coalition on Wastewater-related COVID-19 Research goes so far as to recommend giving wastewater data "the same level of security that groups of individual health information data" are given (CWN, 2020, p. 2). However, it is dubious whether such a high level of security is ethically required for community wastewater surveillance data, which are not individually identifiable. Moreover, requiring a high-level information security architecture introduces barriers to the adoption of wastewater surveillance systems. It also has costs in terms of obstructing the flow of data across users for scientifically useful purposes. For these reasons, data security measures need not rival those for more sensitive health data. Nevertheless, it is important to have reasonable systems in place for preventing unauthorized access.

A second dimension of the obligation to ensure appropriate use of data is preventing "function creep," or expansion of the purposes for which wastewater data are used. To maintain public trust in the wastewater-based infectious disease surveillance system, firewalls must be maintained that prevent transmission of data to aid in law enforcement efforts (Joh, 2021; Ram et al., 2022; Scassa et al., 2022). Infrastructure built for public health surveillance of wastewater (e.g., analytical platforms, periodic water sampling) should not be used by law enforcement officials for their own purposes. Decisions also need to be made about uses of the data by public health officials for purposes other than communicable disease control—for example, estimating substance use in various populations for purposes of allocating programmatic resources (Scassa et al., 2022).

A third dimension of appropriate data use is making wise decisions about data sharing. Prior ethical guidance for wastewater surveillance emphasizes the obligation to share data with other public health agencies (CWN, 2020; Hrudey et al., 2021), and the data sharing obligation arguably extends to academic researchers and others who can assist with analyses, especially during emergencies (Hrudey et al., 2021; WHO, 2017). There is also a strong ethical argument that it extends further, to require sharing of aggregated data with the public (Ram et al., 2022). Members of the public may find such data reports valuable in their efforts to protect themselves—for example, information about SARS-CoV-2 in wastewater in communities with low testing rates allows residents and travelers to take extra precautions (see Chapter 2)—and sharing information learned from wastewater surveillance can help make the benefits of the system clear to the public. Principles of equity and transparency also support sharing findings from wastewater surveillance with key government leaders and the public, accompanied by comprehensible information to help people understand the limitations of the data and the kinds of inferences they can and cannot reliably support.

Whether to share more granular and comprehensive data or actual wastewater samples is a more difficult question. Such data sharing could yield important scientific and public health benefits. On the other hand, the larger the number of users, the greater the risk that misuse of data (e.g., function creep, attempts to identify individuals) will occur, or that data may be analyzed in ways that result in stigmatization of particular communities (Manning and Walton, 2021). Because of these risks, the NWSS does not publicly share even aggregated findings from sewersheds with fewer than 3,000 people;[6] access to more granular data is restricted to public health departments (Naughton et al., 2021). Some liberalization of this policy may be warranted, but requests for sharing of samples and more granular data (beyond what is publicly shared) require careful, case-by-case consideration by an appropriately constituted group of experts (e.g., a data use committee). Particular care should be exercised before deciding to share data or samples from surveillance activities at the sub-community level (e.g., sentinel sites). In small-scale cases, the privacy concerns may well outweigh the benefits of data sharing and an access policy like the NWSS's may be desirable.

Fostering public acceptance necessitates that concerns about data sharing and potential expansions of the scope of wastewater surveillance are addressed and requires good governance, accountability, and transparency (WHO, 2017). These objectives are best pursued through the creation of a decision-making body with public health expertise as well as community representation and statistical and ethics expertise. Such a body is best able to balance the public health mission against countervailing concerns and identify means of minimizing risks that are of concern to the community. For example, statisticians can evaluate the prospects for techniques to minimize risk of identification and the potential effects of proposed linkages to other data sets (Jacobs et al., 2021). Although careful ethical analysis and communication is unlikely to surmount all public concerns, building confidence that an intervention is evidence-based and that officials have heard and considered public concerns about the intervention is essential to the work of building trust and legitimacy (CWN, 2020).

Data use committees (DUCs) have been discussed in the bioethics literature for other situations where secondary uses of data are contemplated and have been adopted in some settings (Scassa et al., 2022). These multi-

[6] See https://nwbe.org/?page_id=77#guidelines.

disciplinary committees evaluate proposed secondary uses of data, assessing ethical concerns and the potential scientific benefits. DUCs provide a useful model for ethical governance of wastewater surveillance systems as well. The history of research ethics suggests that only when oversight structures are in place will ethical norms become an ingrained part of practice (Fairchild and Bayer, 2004). Among the tasks that an ethics committee for wastewater surveillance (modeled after DUCs) could undertake is the development of a carefully crafted, standard data use agreement for academic and industry partners who wish to analyze wastewater data. By specifying ground rules for data users, such agreements can minimize the risk that data sharing will result in reidentification or unauthorized analyses.

The ethics committee's assessment of potential expansions of wastewater data collection or use should anchor on the principles of proportionality, equity, and transparency. Proportionality requires that the public health benefits of an intervention outweigh the burdens and risks—a basic principle of public health ethics. For surveillance efforts, assessing the benefit–risk balance requires ascertaining that the information collected is both useful and actually used and that the burdens on individuals and communities (e.g., stigmatization potential) have been minimized (Lee et al., 2012; WHO, 2017). Minimizing burdens means considering not only whether wastewater surveillance can be performed in a way that reduces risks to individuals and communities but also evaluating whether these burdens and risks are less than other effective methods of disease surveillance. The principle of equity imposes the further requirements that burdens are distributed fairly across communities and that the benefits of wastewater surveillance accrue to the groups whose interests (e.g., privacy) are burdened (Scassa et al., 2022). Finally, the principle of transparency should prompt ethics committees to consider how to explain their recommendations to the public. If the rationale for an expansion in wastewater surveillance cannot be comfortably communicated to the public, that may be a sign that its ethical defensibility is questionable.

## ASSURING DATA QUALITY AND ACTIONABILITY

Another important consideration to achieve the vision for a national wastewater surveillance system involves data quality and actionability. State and local wastewater surveillance programs continue to evolve to better serve both local and national needs, and these programs may look different from one locality to the next. But across all systems, reliable data and scientifically sound interpretation are essential so that they are trusted by public health practitioners and the general public.

### Advancing Data Quality and Comparability

Wastewater surveillance has, by necessity, progressed more rapidly than most other public health surveillance tools; therefore, ongoing efforts are needed to advance data quality and comparability. The COVID-19 pandemic spurred many research laboratories to develop their own analytical methods for wastewater surveillance, including optimizing sample collection and handling, viral concentration, nucleic acid extraction, virus or variant quantification, and computational methods for data analysis. These efforts produced a rich set of resources, but as wastewater surveillance shifts into longer-term and expanded uses, particularly at a national scale, the strengths and weaknesses of different methods need to be assessed and trade-offs rigorously evaluated. Using different methods is reasonable for looking at trends within individual communities, but when implementing a national system, one or a few validated approaches would be best to assure that the data are directly comparable. The vast number of published data sets collected with a diversity of methods will serve as a valuable resource to identify a limited set of the best available methods (e.g., those that best ameliorate wastewater matrix biases and balance quantitative DNA recovery with DNA quality for sequencing).

Even if agreement could be reached about the best available methods, a single standard method may be difficult to implement across a national wastewater surveillance system. Each locality has different technical capabilities and regulatory environments, and localities may have different preferences for use of public health laboratories versus contracted private laboratories. Some methods may be cost-prohibitive for some laboratories (e.g., magnetic bead viral concentration approaches that dramatically increase sensitivity and throughput).

The goal is a national system within which data are comparable across geographic areas for both point estimates of pathogen load and to assess spatial and temporal trends. Some level of methodological flexibility within the national wastewater surveillance system can be afforded as long as expectations for cross-validation with specific samples or standards are established and the data can be transformed by statistical methods into a comparable national-scale data set. Ongoing quality assurance and quality control will be important to provide the data required to correct for specific technical biases introduced by particular methods. Additional effort is needed to compile this information and develop a limited subset of approved methods whose outputs can be statistically transformed into comparable data. The committee's Phase 2 report will discuss methods for sampling and analysis and quality criteria in more detail.

Several challenges stand in the way of achieving standardized wastewater surveillance data that are comparable across geographic areas, beyond differences in analytical methods. Variability in the data across different localities is introduced through factors such as wastewater system characteristics, population mobility, sampling method, sample transit time, and differences in flow normalization methods to account for dilution due to rainfall or other factors (see also Chapter 3) (Stadler et al., 2020; Wade et al., 2022). All of these issues necessitate ongoing research into best practices to continually improve data standardization methods and resources for disseminating these practices in support of a national wastewater surveillance system.

In summary, a national wastewater surveillance system can accommodate some degree of diversity in sampling and analytical approaches but only if additional investments are made in developing methods for standardizing data for purposes of comparison across geographic areas.

## Advancing Data Interpretation and Actionability

In addition to generating reliable and comparable data, a national wastewater surveillance system needs to support timely interpretation of those data to support public health actions. Collecting, analyzing, and statistically interpreting wastewater data on a short turnaround is necessary for the data to be useful. Specifically, "data need to be available within 5–7 days of sample collection to ensure timely application for response decisions" (Kirby et al., 2021). To achieve this, the committee envisions a national data management system that supports rapid data importation from analytical laboratories, quick access to analytical results, easy-to-use data analysis tools appropriate to support the needs of public health decision makers, and public health agency staff who are trained to use the tools and understand the results. Training and capacity building are discussed in more detail in the next section.

Various methods are employed to draw inferences that wastewater data indicate a worsening trend in pathogen load, from tests of statistical significance to application of expert judgment. These analyses require skills that are not always readily available within the jurisdiction or local public health department (McClary-Gutierrez et al., 2021). The NWSS data repository (Data Collation and Integration for Public Health Event Response [DCIPHER]) provides basic analytical tools for states or localities that do not have independent data analytics capacity, although there is room to strengthen and expand available data analysis methods and tools and better integrate with other relevant disease surveillance data. Many localities are developing advances in data analytics, and CDC should learn from these efforts to identify improved approaches to add to DCIPHER and bring these tools to all participating localities. For example, the State of Wisconsin is working with data science researchers to develop benchmarks to better describe an upward trend. Statistical consultants have worked with North Carolina to better discern signal from noise in SARS-CoV-2 trends (Keshaviah et al., 2022a).

Additional research is needed to improve the usefulness and comparability of wastewater surveillance data. Although trend analysis from wastewater surveillance remains a critical interpretation metric for public health monitoring, absolute quantification of the number of infected individuals is not yet firmly established from wastewater surveillance because there have been minimal fecal shedding or epidemiology studies that can help link quantitative SARS-CoV-2 levels to the number of infected individuals. Far fewer data are available for other potential target pathogens for future expanded wastewater surveillance, thus necessitating research and validation studies to establish the basis for interpreting concentration levels of new target pathogens.

In summary, additional investments to accelerate access to wastewater surveillance data and to continue to improve available data analysis methods and robust tools in the NWSS would improve interpretation and enhance the actionability of wastewater surveillance.

## BUILDING BROAD AND SUSTAINABLE CAPACITY

Ensuring that wastewater surveillance data are useful for informing public health action requires not only appropriate methods and tools to generate reliable data and information but also sufficient capacity among the partners that make up the wastewater surveillance system and a reasonable expectation that this capacity can be sustained. This includes having a sufficient number of staff for sampling, analysis, and data interpretation; appropriate equipment to perform the work efficiently and effectively; and funding to sustain the program over time. Specialized training will also be important for laboratory and epidemiology practitioners to become familiar with the approach. Wastewater infectious disease surveillance is a relatively new public health strategy in the United States, necessitating continued capacity development and efforts to increase awareness of its benefits.

### Creating and Maintaining Operational Capacity for Sampling and Analysis

As the NWSS continues, potentially expanding in geographic and pathogen coverage and settling into long-term processes, attention is needed to ensure that the operational capacity within laboratories and wastewater utilities—both staffing and equipment—is appropriately developed and maintained to support the system moving forward.

Participation of wastewater utilities is the backbone of the surveillance system. Although utilities routinely collect inflow samples as part of operational or compliance sampling to ensure proper performance, wastewater surveillance involves stricter sampling requirements and additional efforts to ship the samples to an analytical laboratory. Many utilities began sampling programs as an in-kind service in response to a national crisis. As surveillance becomes routine, the burden of sampling and personnel time needs to be compensated through funding mechanisms. Streamlined mechanisms to provide ongoing funding support for staff, equipment, and other expenses should continue to be pursued to support sustainable capacity. Utilities can contribute valuable insights about system characteristics, such as flow and industrial inputs, to support better data analysis and interpretation, and involving wastewater utilities in the design and refinement of surveillance sampling is also important to gain long-term buy-in. Strong relationships between public health agencies and wastewater utilities will help ensure that these partners remain engaged and prepared if increased surveillance activities are triggered. CDC, EPA, and stakeholder groups such as the Water Environment Federation (WEF) can help support and facilitate these relationships. WEF is also facilitating training and sharing of expertise among utilities through one of the Communities of Practice (see Box 4-1).

Analytical laboratories are another critical partner in wastewater surveillance. Initial development and monitoring for SARS-CoV-2 occurred in research laboratories in many localities, but wastewater surveillance sample analysis is now transitioning to public health laboratories or private laboratories. Essentials for this transition include space, instrumentation, and personnel with specialized wastewater analysis training. Wastewater samples are unlike clinical samples, have a high level of heterogeneity, and often require troubleshooting, even after methods are established. Specific training and experience with sample handling, analysis, and data interpretation are needed. CDC has established a national commercial testing contract to support those without public health laboratory capacity.

Maintaining some level of baseline wastewater infectious disease surveillance, even at a low level, is important to ensure that the institutional framework will be in place to broaden and/or scale up efforts, as needed, in the face of new or re-emerging threats. Ongoing training and succession planning will be important to maintain the expertise.

Increased investment in technology development that covers sampling devices and analytical methods would also strengthen the capacity of a national wastewater surveillance system. The current state of the art in wastewater surveillance has been advanced rapidly by new technologies during the COVID-19 pandemic, but further technological advances can help scale up and generalize deployment. For example, advances in nucleic acid sequencing could reduce cost and turnaround time. Flow-weighted sampling equipment is expensive, fragile, and prone to clogging. As new

**BOX 4-1**
**Wastewater Surveillance Communities of Practice**

In a public health context, Communities of Practice provide a collaborative framework for public health professionals to share communal learning, expertise, and knowledge. Communities of Practice are autonomous and distinct from other groups (e.g., a working group) in that communication at all levels is more transparent and decisions are community-based, focusing equally on three elements:

1. Domain: the shared interest by which the group comes together.
2. Community: the group of individuals brought together by the shared interest.
3. Practice: the ways of addressing issues that are agreed upon by the group.

State, tribal, local, and territorial health departments; public health laboratories; and local wastewater utilities can join one of three respective Communities of Practice dedicated to developing COVID-19 wastewater surveillance with the National Wastewater Surveillance System (NWSS). The NWSS health department community of practice is led by the U.S. Centers for Disease Control and Prevention (CDC), while the laboratory community of practice is led by the Association of Public Health Laboratories, and the wastewater utilities community of practice is led by the Water Environment Federation (WEF). All are funded by CDC. They hold monthly meetings to share information, improve data coordination and interpretation, and support multidisciplinary relationship-building and public health action.

The state health department community of practice members are encouraged to develop local Communities of Practice as well. For example, the Virginia Department of Health Office of Environmental Health Services is a member of CDC's national community of practice and leads a Virginia-based community of practice to specifically support Virginia's wastewater surveillance efforts. The Virginia Community of Practice comprises various groups, including local wastewater and sewer utilities, laboratories, university-based research groups, towns, cities, and others.

SOURCES: https://www.cdc.gov/publichealthgateway/phcommunities/resourcekit/intro/introduction-to-cops.html and https://www.vdh.virginia.gov/content/uploads/sites/20/2021/05/Virginia-WWS-CoP-Slideshare.pdf.

methods continue to emerge, CDC will need to support the development of standards, including reference materials, and approaches for cross-laboratory validation of analytical methods.

## Improving Capacity for Local Interpretation of Wastewater Surveillance Data

A robust wastewater surveillance system requires strong capacity for analyzing and interpreting data in support of decision making at the state, tribal, local, and territorial levels. Currently, local interpretation and use of wastewater surveillance data range widely, and a number of communities collect the samples but do not use the data for local decision making, largely due to limitations in workforce capacity and training. Guidance and training on the analysis and interpretation of wastewater data would help build understanding and confidence in evaluating wastewater data patterns and using that information for public health decisions.

Looking forward toward an expanded NWSS, public health guidance and training will need to include the following:

- How to interpret the implications of wastewater surveillance data for disease epidemiology across a range of pathogens. For example, what does a positive sample for a given infectious disease imply about the likelihood of a case of a disease in the catchment population? What does an increasing trend, concentration of a given level, or other wastewater information imply about the population patterns of disease?
- Explanations of the wastewater surveillance data analysis approaches that would be appropriate for different use cases, including their limitations and when they can and cannot be validly applied.
- Clear, nontechnical guidance on the standards for wastewater surveillance collection and laboratory methods so that public health personnel who may not have laboratory training can understand the potential applications, uses, and limitations of different data sets.
- Explanation of NWSS analytical capabilities, use case examples, and the basis for interpretations provided from nationally based analytical pipelines.

It is not necessary that all public health personnel be trained on all aspects of wastewater surveillance (i.e., methods, data analysis, interpretation, and potential actions). Rather, an individual or smaller group from state, tribal, local, and territorial health departments could participate in such training and serve as a liaison to enable information sharing and coordination within the NWSS and other units within a public health jurisdiction. Trained state personnel can also serve as a resource and provide guidance to local health departments within the state.

One of the largest initial barriers among public health practitioners to using wastewater surveillance data was unfamiliarity (Keshaviah et al., 2022b; McClary-Gutierrez et al., 2021). Seeing how others use these data in their own community raises confidence (McClary-Gutierrez et al., 2021), and CDC Communities of Practice (see Box 4-1) provide opportunities for laboratory personnel and public health practitioners to make connections among peers and share their experiences and lessons learned. Also, the formation of smaller cohorts has been beneficial to build relationships among individual entities and encourage peer-to-peer discussion. Dedicated sessions at professional conferences, such as the Council of State and Territorial Epidemiologists Annual Conference, would increase exposure to community-based wastewater surveillance, which is becoming a highly interdisciplinary specialty. Additionally, the new COVID-19 National Wastewater Surveillance System Centers of Excellence (see Box 4-2) can further support public health practitioners through training and development of best practices.

In summary, training public health practitioners and improving access to those with experience with wastewater surveillance implementation, analysis, and data interpretation is expected to increase the use of wastewater surveillance data for public health action.

### Expanding the Talent Pipeline

Training within university programs in biomedical informatics, biostatistics, environmental engineering, epidemiology, microbiology, or other related fields is necessary to develop and maintain a workforce with the

**BOX 4-2**
**COVID-19 Wastewater Surveillance Centers of Excellence**

In August 2022, the Centers for Disease Control and Prevention (CDC) named two National Wastewater Surveillance System (NWSS) Centers of Excellence: (1) the Colorado Department of Public Health and Environment in collaboration with the University of Denver; and (2) the Houston Wastewater Epidemiology initiative, which is a collaboration among the Houston Health Department, Houston Public Works, and Rice University. The Centers will serve multiple functions within two broad themes: (1) providing expertise for training, implementing, and communicating; and (2) further developing, validating, and troubleshooting data generation and analysis for the NWSS. Drawing from their experiences implementing wastewater surveillance—both joined the NWSS in September 2020—they will train other localities to conduct wastewater surveillance, including sampling, analysis, and interpretation for public health action, while helping NWSS participants maintain connections with the latest technology and research. Technology transfer of methods for new pathogen targets will also be an important function of the Centers. With the volumes of data that are being generated, projects within Centers of Excellence can focus on retrospective analysis to mine data for trends and define important benchmarks that correlate to key community health indicators.

SOURCES: https://news.rice.edu/news/2022/cdc-names-houston-health-department-rice-wastewater-epidemiology-center-excellence and https://nwbe.org/?page_id=77.

qualifications needed to continuously improve and execute wastewater surveillance in public health laboratories and departments. For example, laboratories could coordinate with academic programs to provide opportunities for internships to support the talent pipeline of students moving into careers in applied public health. Three workforce gaps that urgently need to be filled are in the areas of epidemiology, health informatics, and bioinformatics.

Within the public health system, one of the major limiting factors for capacity building is recruiting and retaining skilled epidemiologists. Wastewater surveillance requires analyzing novel, high-dimensional data and developing completely new analytic approaches. The skills required go far beyond those of most entry-level epidemiologists. Highly trained, PhD-level epidemiologists are currently in short supply due to the increase in demand during the COVID-19 pandemic. Cross-training to include environmental engineering aspects of wastewater surveillance would be particularly valuable.

Robust health informatics support that can set up and maintain complex systems is needed for automating and disseminating wastewater data efficiently. A wide range of health informatics is involved in wastewater surveillance that, ideally, would be integrated with other data streams within the public health system. Automating data streams and reporting is essential for long-term sustainability and efficiency. Informatics expertise is also needed for creating dashboard interfaces and automating customizable reports to be meaningful to various stakeholders, including wastewater utilities, local health departments, and policy makers.

Bioinformatics expertise for sequence analysis is essential to analyze and interpret complex sequence data. Bioinformatics training needs for wastewater surveillance are distinct from expertise in most clinical sequencing because wastewater contains a composite mixture of different strains and variants of targets of interest rather than the single strain typically found in a given clinical sample.

### Predictable and Sustained Funding

Wastewater surveillance is in its infancy as a field, and to date, CDC funding has been critical to the advancement of the field and establishment of the NWSS. To maintain and advance wastewater surveillance and retain participation of states, tribes, territories, and localities in the NWSS, it is critical that sustained, predictable federal funding remain available. Predictable and sustained CDC funding is needed in two key areas:

1. to support capacity building (e.g., trained staff, appropriate equipment) as the system transitions from emergency pandemic response to a sustained program; and

2. advances in science and technology so that the NWSS can continue to improve its capability and address additional organisms, including emerging pathogens, for which limited information may be available (see Figure 4-1).

For example, technology advancements and process improvements could include new methods for sample concentration and new algorithms for analyzing data. Long-term research investments are also needed to adapt methods for evolving SARS-CoV-2 variants, develop and validate new targets, and realize the full potential of wastewater surveillance. These technology development and science needs span topics traditionally funded by different government agencies (e.g., National Institutes of Health [NIH], National Science Foundation [NSF], National Institute of Standards and Technology [NIST], EPA, U.S. Food and Drug Administration, U.S. Department of Agriculture (USDA), and National Oceanic and Atmospheric Administration), private sources (e.g., Rockefeller Foundation, Alfred P. Sloan Foundation), and companies. Unpredictable funding makes it highly challenging to retain laboratory personnel, informaticians, and public health epidemiologists who are vital to maintaining a surveillance system and to recruit students, researchers, and companies to participate in these efforts.

## ACHIEVING INTEGRATION AND COLLABORATION

Achieving the vision of an integrated, actionable wastewater surveillance program requires coordination and collaboration across many par-

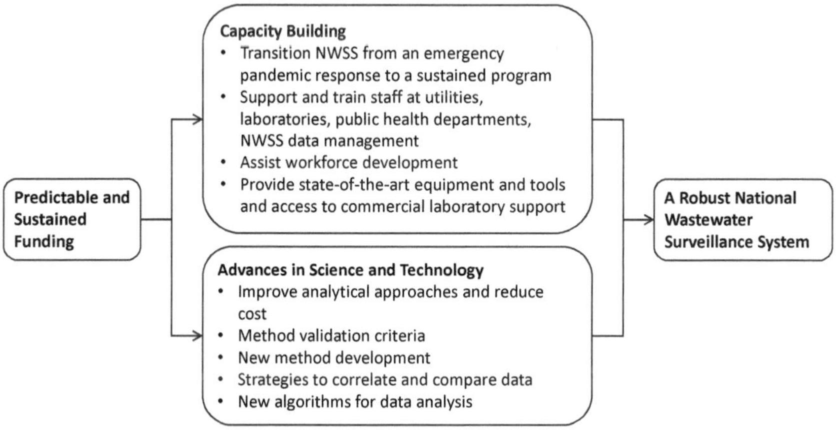


**FIGURE 4-1** Areas for sustained investment to support the National Wastewater Surveillance System (NWSS).

ties within the public health system, including local public health agencies, CDC, analytical laboratories, and wastewater utilities. A coordination strategy will be unique to the individual state, tribe, locality, or territory, but clear designation of roles (e.g., sample collection, laboratory analysis, data interpretation and integration, and data dissemination) will be necessary. Typically, the primary public health agency[7] overseeing the wastewater surveillance program provides data interpretations and ensures that the wastewater data are integrated with other data sources (e.g., clinical testing, syndromic data) to inform sound decision making. This effort requires not only coordination with laboratories whose staff understand methodological constraints and utilities that provide analytical context and important expertise for working with the complex matrix of wastewater but also coordination *within* public health jurisdictions, including epidemiologists and practitioners who ultimately use the data to decide upon public health actions. Furthermore, to build out the system, innovation and progressive improvement of analytical methods, sampling approaches, and data analysis tools and methods are needed, which often fall upon academics and other entities within the scientific community that have expertise in moving analytical techniques and data analysis approaches from discovery to implementation.

### Coordination Within Public Health Jurisdictions

Early adopters of wastewater surveillance have developed helpful models for coordination within public health jurisdictions to better support decision making, although the optimal approach may vary across localities. Some jurisdictions are small and lack capacity to undertake new activities, deferring to state public health systems to implement wastewater surveillance, while large municipalities or counties may operate their own surveillance systems. Typically, a primary lead wastewater surveillance group is established within a larger public health agency to compile and interpret data and actively engage partners (see Figure 4-2). Practitioners within these units are often familiar with the complexities of environmental samples and with epidemiology and community health interventions. These units often serve as liaisons between the analytical laboratory and other components of a public health agency or system (e.g., infectious diseases, epidemiology). This lead group can help bridge communication gaps between highly technical environmental engineering or laboratory information and epidemiological considerations and public health actions

[7] For the purposes of this report, primary is considered to be one of the 64 state, territorial, city, or county health departments eligible for Epidemiology and Laboratory Capacity (ELC) funding (see also Chapter 1).

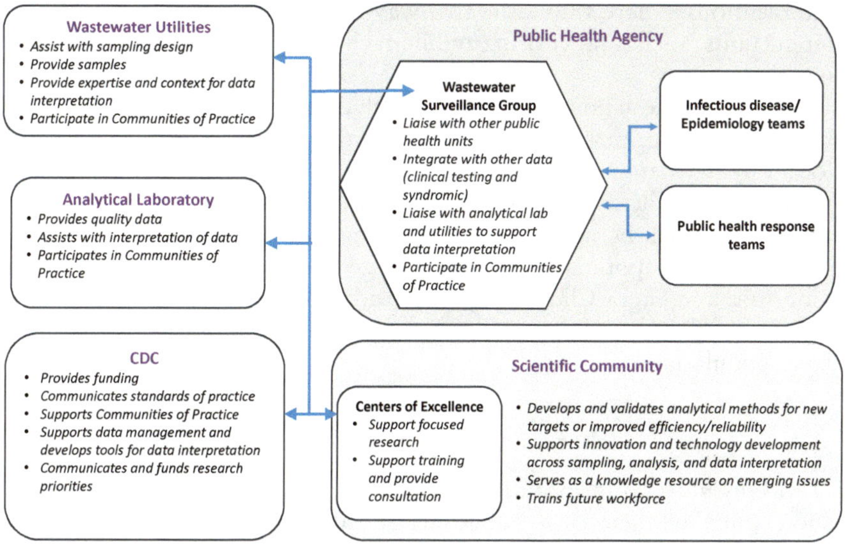


**FIGURE 4-2** Framework for wastewater surveillance coordination and collaboration within the public health system. The public health agency's wastewater surveillance group collaborates with wastewater utilities and laboratories, which play a critical role in providing samples, data, and expertise, and also leads internal coordination with infectious disease/epidemiology teams to integrate wastewater data into public health actions. The public health agency typically represents the primary organization receiving ELC funding (e.g., states, large cities/counties) but may also include other large cities and counties with sufficient staff capacity for data analysis and coordination. State public health agencies typically also coordinate with local public health departments, which serve as the frontline of public health response, The scientific community is critical for advancing wastewater surveillance for new technologies, research, and training.

in infectious disease. Technical expertise needed within a core, lead wastewater surveillance group includes epidemiology, environmental engineering, and sound technical knowledge for laboratory analysis. Certain aspects of the expertise could be gained from engaging team members from outside of the unit, such as from wastewater utilities or the analytical laboratory.

It is critical that the environmental and communicable disease arms of the public health system have close coordination with wastewater surveillance teams. For example, an organization may include the same epidemiologists who are involved in wastewater surveillance in the agency's overall epidemiology response teams. Public health practitioners, who are ultimately the end users of the data, have the contextual knowledge to drive the data needs to identify how wastewater data can best inform interventions. There should be two-way communication between the different units, including crafting public-facing messaging with the communications teams.

To maximize actionability, local health departments need to be engaged, because they can provide input on community characteristics and ensure local buy-in. State public health systems that lead wastewater surveillance efforts are excellent points of contact for local and county public health officials and will be most familiar with the state's specific needs, barriers, and vulnerable populations. Many large municipalities (e.g., Chicago, Houston, and Santa Clara County) already have robust public health systems and operate sophisticated wastewater surveillance teams that cover large populations.

### Collaboration with Analytical Laboratories

Many states and localities are implementing wastewater surveillance successfully in their own public health laboratories, which are centralized and have the necessary equipment. The response to the COVID-19 pandemic spurred improvements in analytical capacity, and training and retention of key personnel is needed to create institutional knowledge. This laboratory expertise should be integrated into public health wastewater surveillance teams. However, some states, tribes, localities, and territories may choose to use academic, commercial, or utilities-based laboratories rather than building laboratory capacity within the public health system. Regardless of whether the laboratory capacity is within the public health agency or external, coordination is essential to understand issues with data comparability due to methodological differences, instrumentation, or other factors. Close communication, such as biweekly calls that include representatives of the public health agency wastewater surveillance group and analytical laboratory, allow for relatively early reactions to data anomalies, and ongoing interactions provide opportunities to understand methodological influences on data trends over longer periods of time. Good communication between the laboratory and the wastewater facilities is also important to obtain necessary parameters for samples and understand the system and representativeness of the samples.

### Collaboration with Wastewater Utilities

Wastewater utilities have knowledge critical to interpreting data that includes sample representativeness, sewershed boundaries, and sample interferences and complications due to operational procedures; therefore, they need to be engaged as full partners. Because the samples themselves are not uniform, wastewater expertise is especially needed to understand uncertainty and variability in the data. Coordination with utilities for sampling often occurs on the local level because local public health entities and wastewater utilities may already interact; however, when such relationships are lacking, state- or national-level assistance with coordination may be needed. National utility member associations, like the WEF, have utility Communities of Practice that can forge and maintain public health/utility partnerships, and facilitate communication among other Communities of Practice and health agencies. Wastewater utilities ideally should interact with both the laboratory and the wastewater surveillance group within the public health agency because their expertise would be important to accurate interpretation of the data. The lead public health wastewater group should proactively invite wastewater utilities to participate beyond providing samples at the level that they are able given personnel and other capacity constraints.

Additionally, some large utilities have research groups or extensive laboratory expertise that would be highly beneficial to draw upon for operating a wastewater surveillance program. For example, Los Angeles Sanitation District and Hampton Roads Sanitation District began wastewater programs early in the COVID-19 pandemic and were able to provide guidance to others in the field.

### Coordination and Collaboration with CDC and Other Federal Partners

Federal-level coordination through CDC to harness the strengths of other federal partners is essential to streamline data curation and visualization, coordinate new method implementation, facilitate training for public health practitioners, conduct public outreach, and provide predictable and sustainable funding. As the surveillance system expands beyond SARS-CoV-2, CDC—the nation's health protection agency—is the logical entity to coordinate the process for the selection of new agents, including working with outside experts who can advise on targets, sampling schemes, data presentation, and interpretation. For emerging diseases and pathogens with zoonotic potential, close coordination with animal health agencies and laboratories (e.g., USDA and the National Animal Health Laboratory Network), is essential to ensure informed and coordinated investigations and response activities.

One major function of CDC's NWSS program will be to coordinate and maintain the data repository and support analytical and visualization tools that are available to national, state, tribal, local, and territorial public health agencies as well as the publicly accessible dashboard. A uniform yet flexible data framework will enable comparisons within and across states, while capturing as many reliable data sources as possible. Data analysis tools provided within this framework (with supporting guidance) will make public health actions more accessible for states, tribes, localities, and territories that do not necessarily have the internal expertise to contextualize wastewater data within other public health surveillance efforts.

CDC currently supports and coordinates Communities of Practice (see Box 4-1), which provide valuable information sharing within individual communities for laboratories, public health practitioners, and wastewater utilities. These communities will be important to achieve the vision of a national wastewater surveillance system as it expands to new areas and new pathogens. CDC can strengthen these efforts by coordinating proficiency and round-robin testing (i.e., exchanging samples for comparisons), and engaging academic or industry expertise as needed to interact with the localities performing surveillance. CDC is currently funding two Centers of Excellence (see Box 4-2) that can provide training in both data generation and data interpretation, develop model communications such as reporting formats and outreach materials, provide rapid devolvement and validation of new targets, and backfill critical knowledge gaps in methods and surveillance approaches.

Another important role for federal agencies is to identify and support high-priority research needs associated with national wastewater surveillance. Such investments solidify the knowledge base on which wastewater surveillance is built and promote ongoing advancements in the field that ultimately will enable more informed public health actions. For example, research is needed to fill knowledge gaps regarding the dynamics of SARS-CoV-2 from infected individuals through wastewater systems as well as to understand the epidemiological links between the numbers of infected persons and wastewater measurements. CDC should coordinate with other relevant federal agencies (e.g., NSF, NIH, EPA, and U.S. Geological Survey) to identify and fund high-priority research. Coordination with NIST for creating resources of readily available standards will be important to support comparability across the national system and assurance of good data.

Finally, perhaps the most critical function of the federal government is to provide predictable and sustained funding that supports equitable coverage across a national system.

### Collaboration with the Broader Scientific Community

Early in the COVID-19 pandemic, university laboratories and water or wastewater utilities drove much of the innovation and rapid implementation of SARS-CoV-2 surveillance and, in doing so, established critically important partnerships (Hoar et al., In Press). Research laboratories are now transitioning away from bulk processing and analysis of samples that now can be analyzed through public health or private analytical laboratories. Looking forward, research laboratories, located in a diversity of health, engineering, and environmental science units both within and outside of academia, can provide an avenue for rapid method development for new pathogen targets, improvement of assays for low-level targets in wastewater, and advancement of modeling and data analysis for current and emerging microbial threats. Existing and new partnerships among the scientific community; analytical laboratories; wastewater utilities; and local, state, tribal, territorial, and national public health agencies provide mechanisms for identifying challenges that could be addressed through research and innovation as well as rapid translation of scientific advances into operations.

The formal designation of two Centers of Excellence (see Box 4-1), the Houston NWSS and the Colorado Department of Public Health and Environment/University of Denver, reflects the importance CDC is placing on external partners within the scientific community. These Centers of Excellence are expected to strengthen connections between the NWSS and the experts in the scientific community, but they cannot cover all the research needs of the NWSS. To support useful expansion of the NWSS beyond SARS-CoV-2, basic discovery and developmental research will be needed to address gaps in understanding. This research includes assessments of whether potential pathogens of concern are detectable and stable in wastewater, and development of sample preservation and analysis methods. In the longer term (5–10 years), new cost-effective methods need to be developed for untargeted pathogen biomarker detection to ensure rapid response to emerging pathogens. Additionally, rapid approaches for adapting target-specific methods to new pathogens would strengthen the response capacity to emerging microbial threats. This will require research into the fundamental chemistry, physics, and microfabrication of novel sequencing systems, likely as a mixture of research at academic institutions, national laboratories, and companies, casting a broad net for new high-risk, high-reward approaches. Phase 2 of the committee's study will examine research needs in detail. CDC should support targeted research and innovation through funding mechanisms such as the CDC Broad Agency Announcements.

In addition to its role in scientific discovery and development, academia will have a key role in workforce training (Hoar et al., 2022). These workforce needs will include the full spectrum of wastewater surveillance from sampling methods and design to laboratory analysis, data analysis, data interpretation, and communication. Understanding the current and future workforce needs will be a critical element in planning. A workforce assessment for the NWSS could identify specific training needs that can be met by community colleges as well as by research universities.

## CONCLUSIONS AND RECOMMENDATIONS

**CDC should develop an open and transparent process for prioritizing targets for wastewater surveillance.** Selecting future targets for wastewater surveillance is a challenging endeavor that balances potential health benefits against resource investments and the capabilities of existing technology. CDC would benefit from an independent external advisory panel, with representation from industry, academia, and public health, to provide periodic guidance and input to this process and ensure that the latest advancements in science and technology are considered. The external advisory panel could also provide rapid consultation in future pandemic emergencies. Public input to the process is important because the community should have the opportunity to have concerns heard and considered before a final plan is implemented.

**Although the committee judges that the benefits of responsibly managed wastewater surveillance outweigh the associated ethical concerns, CDC should address privacy concerns through clear public communication and by convening an ethics advisory committee.** CDC should develop and disseminate additional public communications designed to inform the public about the data generated in wastewater surveillance and how these data are used. In addition, CDC should empanel a standing ethics advisory committee to recommend guidelines about the conditions under which wastewater data may be shared with others and to evaluate future expansions of data collection and data access. It is desirable for academic and industry partners to be able to conduct and contribute analyses of wastewater data, which requires responsible data sharing. The ethics committee, which could be modeled after existing data use committees, should create a formal process for executing data use agreements to help address privacy concerns and alleviate burdens in managing data sharing at a local level. Furthermore, if the prospects for identifying individuals in wastewater data strengthen over time, or if any agency or private-sector organization expresses interest in using wastewater data for purposes other than infectious disease surveillance, this body should re-evaluate the balance of health benefits versus risks associated with data sharing and any proposed expansions in data collection and data linkage. There should be a strong firewall maintained that precludes use of data by law enforcement. In performing its work, the ethics body should consider whether steps are needed to help avoid stigmatization of particular communities or to build further buy-in to wastewater surveillance among members of particular communities.

**The effectiveness of the NWSS will depend on predictable and sustained federal investments.** The COVID-19 pandemic emergency spurred many researchers and utilities to volunteer their labor and donate resources in support of the effort, but the vision of a sustained national wastewater surveillance system necessitates a shift from volunteerism to a strategic national plan with well-defined roles supported by federal investments. Federal funding is needed to continue to advance sampling and analysis methods and data analysis tools to improve data quality, comparability, and actionability. Predictable funding is also essential to maintain the workforce capacity and institutional knowledge to sustain a well-functioning wastewater surveillance system that is useful to public health agencies and to support an effective system for data management and interpretation for all public health agencies.

**Close coordination among public health agencies, analytical laboratories, and wastewater utilities is essential to generate reliable data and support appropriate data interpretation and use.** CDC's Communities of Practice for wastewater utilities, laboratory personnel, and public health practitioners provide valuable support for coordination within each of these fields, and CDC can work with these communities to establish expectations for coordination and collaboration with other agencies. State, tribal, local, and territorial public health agencies should also work to strengthen relationships across these partners—for example, by encouraging biweekly meetings with staff from the public health agency, the analytical laboratory department, and the wastewater utility, as appropriate, in support of data interpretation. CDC, as the nation's health protection agency, should continue to lead the coordination of the many federal partners in support of this effort.

**Because the function of the NWSS depends on the participation of wastewater utilities, CDC and local public health agencies should continue to strengthen relationships with wastewater partners.** CDC should continue to work to improve the connections between wastewater utilities and local, state, tribal, territorial, and federal public health agencies, beyond what is currently provided in the Communities of Practice. At a federal level, CDC could set expectations and standards of practice that utilities be engaged as full partners, with compensation for their participation and education and data sharing to ensure that the utilities see the value of their contributions. Local public health agencies should work to build relationships with utilities, who can also provide important expertise essential to developing sound sampling designs and accurate data interpretation.

**Looking forward, academia and the broader scientific community are essential to drive innovation in sampling, laboratory analysis, data management and interpretation, and public communication.** CDC is commended for launching two initial Centers of Excellence, which will help support targeted research and training. In addition to the Centers of Excellence, CDC should engage the scientific community around specific sampling, analytical, and data management needs through funding mechanisms such as the CDC Broad Agency Announcements. Academic and other research laboratories could provide needed training, and an NWSS workforce needs study would help ensure that a trained workforce can meet current and future needs.

# References


Aarestrup, F. M., and M. E. J. Woolhouse. 2020. Using sewage for surveillance of antimicrobial resistance. *Science* 367(6478):630–632. https://doi.org/10.1126/science.aba3432.

Abe, M., H. Katano, M. Nagi, Y. Higashi, Y. Sato, K. Kikuchi, H. Hasegawa, and Y. Miyazaki. 2020. Potency of gastrointestinal colonization and virulence of *Candida auris* in a murine endogenous candidiasis. *PLOS One* 15(12):e0243223. https://doi.org/10.1371/journal.pone.0243223.

Adhikari, S., and R. U. Halden. 2022. Opportunities and limits of wastewater-based epidemiology for tracking global health and attainment of UN Sustainable Development Goals. *Environment International* 163:107217. https://doi.org/10.1016/j.envint.2022.107217.

Adu, F. D., F. A. Kembi, A. Bamgboye, and M. Osei-Kwasi. 1998. Wild polio virus surveillance in the sewage system of selected communities at the risk of poliomyelitis in southwest Nigeria. *East African Medical Journal* 75(2):97–99.

Agrawal, S., L. Orschler, S. Tavazzi, R. Greither, B. M. Gawlik, and S. Lackner. 2022. Genome sequencing of wastewater confirms the arrival of the SARS-CoV-2 Omicron variant at Frankfurt Airport but limited spread in the city of Frankfurt, Germany, in November 2021. *Microbiology Resource Announcements* 11(2):e0122921. https://doi.org/10.1128/MRA.01229-21.

Ahmed, W., N. Angel, J. Edson, K. Bibby, A. Bivins, J. W. O'Brien, P. M. Choi, M. Kitajima, S. L. Simpson, J. Li, B. Tscharke, R. Verhagen, W. J. M. Smith, J. Zaugg, L. Dierens, P. Hugenholtz, K. V. Thomas, and J. F. Mueller. 2020. First confirmed detection of SARS-CoV-2 in untreated wastewater in Australia: A proof of concept for the wastewater surveillance of COVID-19 in the community. *Science of the Total Environment* 728:138764. https://doi.org/10.1016/j.scitotenv.2020.138764.

Ahmed, W., A. Bivins, S. Metcalfe, W. J. M. Smith, M. E. Verbyla, E. M. Symonds, and S. L. Simpson. 2022. Evaluation of process limit of detection and quantification variation of SARS-CoV-2 RT-qPCR and RT-dPCR assays for wastewater surveillance. *Water Research* 213:118132. https://doi.org/10.1016/j.watres.2022.118132.

Alexander, D. J. 2007. An overview of the epidemiology of avian influenza. *Vaccine* 25(30):5637–5644. https://doi.org/10.1016/j.vaccine.2006.10.051.

AMC (Antimicrobial Resistance Collaborators). 2022. Global burden of bacterial antimicrobial resistance in 2019: A systematic analysis. *The Lancet* 399:629–655. https://doi.org/10.1016/S0140-6736(21)02724-0.

Anderson, R. M., and R. M. May. 1992. *Infectious diseases of humans: Dynamics and control.* Oxford: Oxford University Press.

ASPR (Administration for Strategic Preparedness and Response) and NSC (National Security Council). 2018. United States Health Security National Action Plan: Strengthening implementation of the international health regulations based on the 2016 Joint External Evaluation. https://www.phe.gov/Preparedness/international/Documents/jee-nap-508.pdf

Barber, G. 2020. One way to potentially track Covid-19? Sewage. *Wired*, April 7. https://wired.com/story/one-way-to-potentially-track-covid-19-sewage-surveillance/.

Bhardwaj, S. K., N. Bhardwaj, V. Kumar, D. Bhatt, A. Azzouz, J. Bhaumik, K. H. Kim, and A. Deep. 2021. Recent progress in nanomaterial-based sensing of airborne viral and bacterial pathogens. *Environment International* 146:106183.

Boehm, A., and E. A. U. Riley. 2022. Wastewater provides a solution for monitoring Omicron's spread. *The Hill*, December 9. https://thehill.com/opinion/healthcare/584894-wastewater-provides-a-solution-for-monitoring-omicrons-spread/.

Borchardt, M. A., A. B. Boehm, M. Salit, S. K. Spencer, K. R. Wigginton, and R. T. Noble. 2021. The Environmental Microbiology Minimum Information (EMMI) Guidelines: qPCR and dPCR quality and reporting for environmental microbiology. *Environmental Science & Technology* 55(15):10210–10223. https://doi.org/10.1021/acs.est.1c01767.

Böttiger, M., and E. Herrström. 1992. Isolation of polioviruses from sewage and their characteristics: Experience over two decades in Sweden. *Scandinavian Journal of Infectious Diseases* 24(2):151–155. https://doi.org/10.3109/00365549209052605.

Brinch, C., P. Leekitcharoenphon, A. S. R. Duarte, C. A. Svendsen, J. D. Jensen, and F. M. Aarestrup. 2020. Long-term temporal stability of the resistome in sewage from Copenhagen. *mSystems* 5(5):e00841–20. https://doi.org/10.1128/mSystems.00841-20.

CCHD (Carteret County Health Department) and NCDHHS (North Carolina Department of Health and Human Services). 2021. COVID-19 monitoring detects elevated levels of COVID-19 in wastewater. https://www.beaufortnc.org/sites/default/files/fileattachments/public_utilities/page/10091/wastewater_joint_release_.pdf.

CDC (U.S. Centers for Disease Control and Prevention). 2019. *Antibiotic resistance threats in the United States, 2019*. Atlanta, GA: Centers for Disease Control and Prevention. http://dx.doi.org/10.15620/cdc:82532.

Cheng, Q., P. A. Collender, A. K. Heaney, X. Li, R. Dasan, C. Li, J. A. Lewnard, J. L. Zelner, S. Liang, H. H. Chang, L. A. Waller, B. A. Lopman, C. Yang, and J. V. Remais. 2020. The DIOS framework for optimizing infectious disease surveillance: Numerical methods for simulation and multi-objective optimization of surveillance network architectures. *PLOS Computational Biology* 16(12):e1008477. https://doi.org/10.1371/journal.pcbi.1008477.

Cheng, Y. C., S. Hannaoui, T. R. John, S. Dudas, S. Czub, and S. Gilch. 2016. Early and non-invasive detection of chronic wasting disease prions in elk feces by real-time quaking induced conversion. *PLOS One* 11(11):e0166187. https://doi.org/10.1371/journal.pone.0166187.

Chik, A. H. S., M. B. Glier, M. Servos, C. S. Mangat, X. L. Pang, Y. Qiu, P. M. D'Aoust, J. B. Burnet, R. Delatolla, S. Dorner, Q. Geng, J. P. Giesy Jr., R. M. McKay, M. R. Mulvey, N. Prystajecky, N. Srikanthan, Y. Xie, B. Conant, S. E. Hrudey, and Canadian SARS-CoV-2 Inter-Laboratory Consortium. 2021. Comparison of approaches to quantify SARS-CoV-2 in wastewater using RT-qPCR: Results and implications from a collaborative inter-laboratory study in Canada. *Journal of Environmental Sciences (China)* 107:218–229. https://doi.org/10.1016/j.jes.2021.01.029.

Cipolletta, S., G. R. Andreghetti, and G. Mioni. 2022. Risk perception towards COVID-19: A systematic review and qualitative synthesis. *International Journal of Environmental Research and Public Health* 19(8):4649. https://doi.org/10.3390/ijerph19084649.

Clason, L. 2022. COVID-19 wastewater efforts confront long-term questions. *Roll Call*, May 16. https://rollcall.com/2022/05/16/covid-19-wastewater-efforts-confront-long-term-questions/.

Crank, K., W. Chen, A. Bivins, S. Lowry, and K. Bibby. 2022. Contribution of SARS-CoV-2 RNA shedding routes to RNA loads in wastewater. *Science of the Total Environment* 806:150376.

Crits-Christoph, A., R. S. Kantor, M. R. Whitney, O. N. Olm, B. Al-Shaye, Y. C. Lou, A. Flamholz, L. C. Kennedy, H. Greenwald, A. Hinkle, and J. Hetzel. 2021. Genome sequencing of sewage detects regionally prevalent SARS-CoV-2 variants. *Mbio* 12(1):e02703–e02720. https://doi.org/10.1128%2FmBio.02703-20.

CWN (Canadian Water Network). 2020. *Ethics and communications guidance for wastewater surveillance to inform public health decision-making about COVID-19*. Canadian Coalition on Wastewater-related COVID-19 Research. https://cwn-rce.ca/wp-content/uploads/COVID19-Wastewater-Coalition-Ethics-and-Communications-Guidance-v4-Sept-2020.pdf.

Dadras, O., S. SeyedAlinaghi, A. Karimi, A. Shojaei, A. Amiri, S. Mahdiabadi, A. Fakhfouri, A. Razi, H. Mojdeganlou, P. Mojdeganlou, A. Barzegary, Z. Pashaei, A. M. Afsahi, P. Shobeiri, and E. Mehraeen. 2022. COVID-19 vaccines' protection over time and the need for booster doses: A systematic review. *Archives of Academic Emergency Medicine* 10(1):e53. https://doi.org/10.22037/aaem.v10i1.1582.

D'Aoust, P. M., T. E. Graber, E. Mercier, D. Montpetit, I. Alexandrov, N. Neault, A. T. Baig, J. Mayne, X. Zhang, T. Alain, M. R. Servos, N. Srikanthan, M. MacKenzie, D. Figeys, D. Manuel, P. Juni, A. E. MacKenzie, and R. Delatolla. 2021. Catching a resurgence: Increase in SARS-CoV-2 viral RNA identified in wastewater 48 h before COVID-19 clinical tests and 96 h before hospitalizations. *Science of the Total Environment* 770:145319. https://doi.org/10.1016/j.scitotenv.2021.145319.

De Jong, E. 2021. Auckland lockdown to end despite three new cases of COVID-19. *The Guardian*, February 17. https://www.theguardian.com/world/2021/feb/17/new-zealand-reports-two-new-cases-of-covid-19-community-transmission.

De Jonge, E. F., C. M. Peterse, J. M. Koelewijn, A. M. R. van der Drift, R. F. van der Beek, E. Nagelkerke, and W. J. Lodder. 2022. The detection of monkeypox virus DNA in wastewater samples in the Netherlands. *Science of the Total Environment* 852:158265. https://doi.org/10.1016/j.scitotenv.2022.158265.

Delamater, P. L., E. J. Street, T. F. Leslie, Y. T. Yang, and K. H. Jacobsen. 2019. Complexity of the basic reproduction number ($R_0$). *Emerging Infectious Diseases* 25(1):1–4. https://doi.org/10.3201/eid2501.171901.

Dixon, B. E., S. Dearth, T. J. Duszynski, and S. J. Grannis. 2022. Dashboards are trendy, visible components of data management in public health: Sustaining their use after the pandemic requires a broader view. *American Journal of Public Health* 112(6):900–903. https://doi.org/10.2105/AJPH.2022.306849.

Eaton, C., S. Coxon, I. Pattis, and B. Gilpin. 2021. Wastewater-based epidemiology: A framework to identify and prioritise health determinants for wastewater monitoring. https://www.esr.cri.nz/assets/files/ Framework-to-identify-September-2021.pdf.

EclinicalMedicine. 2021. Antimicrobial resistance: A top ten global public health threat. *EclinicalMedicine* 41:101221. https://doi.org/10.1016/j.eclinm.2021.101221.

Elliot, P., O. Eales, N. Steyn, D. Tang, B. Bodinier, H. Wang, J. Elliott, M. Whitaker, C. Atchison, P. J. Diggle, A. J. Page, A. J. Trotter, D. Ashby, W. Barclay, G. Taylor, H. Ward, A. Darzi, G. S. Cooke, C. A. Donnelly, and M. Chadeau-Hyam. 2022. Twin peaks: The Omicron SARS-CoV-2 BA.1 and BA.2 epidemics in England. *Science* 376(6600). https://doi.org/10.1126/science.abq4411.

EPA (U.S. Environmental Protection Agency). 2021. *A compendium of U.S. wastewater surveillance to support COVID-19 public health response*. https://www.epa.gov/system/files/documents/2021-09/wastewater-surveillance-compendium.pdf.

Erster, O., I. Bar-Or, V. Levy, R. Shatzman-Steuerman, D. Sofer, L. Weiss, R. Vasserman, I. S. Fratty, K. Kestin, M. Elul, N. Levi, R. Alkrenawi, E. Mendelson, M. Mandelboim, and M. Weil. 2022. Monitoring of enterovirus D68 outbreak in Israel by a parallel clinical and wastewater based surveillance. *Viruses* 14(5):1010. https://doi.org/10.3390/v14051010.

Fairchild, A. L., and R. Bayer. 2004. Public health: Ethics and the conduct of public health surveillance. *Science* 303(5658):631–632. https://doi.org/10.1126/science.1094038.

Fairchild, G., P. M. Polgreen, E. Foster, G. Rushton, and A. M. Segre. 2013. How many suffice? A computational framework for sizing sentinel surveillance networks. *International Journal of Health Geographics* 12(1):56. https://doi.org/10.1186/1476-072X-12-56.

Farkas, K., C. Pellett, N. Alex-Sanders, M. T. P. Bridgman, A. Corbishley, J. M. S. Grimsley, B. Kasprzyk-Hordern, J. L. Kevill, I. Pântea, I. S. Richardson-O'Neill, K. Lambert-Slosarska, N. Woodhall, and D. L. Jones. 2022. Comparative assessment of filtration- and precipitation-based methods for the concentration of SARS-CoV-2 and other viruses from wastewater. *Microbiology Spectrum* 10(4):e0110222. https://doi.org/10.1128/spectrum.01102-22.

Feng, S., A. Roguet, J. S. McClary-Gutierrez, R. J. Newton, N. Kloczko, J. G. Meiman, and S. L. McLellan. 2021. Evaluation of sampling, analysis, and normalization methods for SARS-CoV-2 concentrations in wastewater to assess COVID-19 burdens in Wisconsin communities. *Environmental Science: Water Research & Technology* 1(8):1955–1965.

Fernandez-Cassi, X., A. Scheidegger, C. Banziger, F. Cariti, A. Tunas Corzon, P. Ganesanandamoorthy, J. C. Lemaitre, C. Ort, T. R. Julian, and T. Kohn. 2021. Wastewater monitoring outperforms case numbers as a tool to track COVID-19 incidence dynamics when test positivity rates are high. *Water Research* 200:117252. https://doi.org/10.1016/j.watres.2021.117252.

Gable, L., N. Ram, and J. L. Ram. 2020. Legal and ethical implications of wastewater monitoring of SARS-CoV-2 for COVID-19 surveillance. *Journal of Law and the Biosciences* 7(1):lsaa039. https://doi.org/10.1093/jlb/lsaa039.

Galani, A., R. Aalizadeh, M. Kostakis, A. Markou, N. Alygizakis, T. Lytras, P. G. Adamopoulos, J. Peccia, D. C. Thompson, A. Kontou, A. Karagiannidis, E. S. Lianidou, M. Avgeris, D. Paraskevis, S. Tsiodras, A. Scorilas, V. Vasiliou, M. A. Dimopoulos, and N. S. Thomaidis. 2022. SARS-CoV-2 wastewater surveillance data can predict hospitalizations and ICU admissions. *Science of the Total Environment* 804:150151. https://doi.org/10.1016/j.scitotenv.2021.150151.

Gershy-Damet, G. M., A. Lanusse, and M. Dosso. 1987. Surveillance des enterovirus dans les eaux usées en Cote-d'Ivoire [Surveillance of enteroviruses in the waste water of the Ivory Coast]. *Bulletin de la Société de Pathologie Exotique et de Ses Filiales* 80(2):180–186.

Graham, K. E., S. K. Loeb, M. K. Wolfe, D. Catoe, N. Sinnott-Armstrong, S. Kim, K. M. Yamahara, L. M. Sassoubre, L. M. Mendoza Grijalva, L. Roldan-Hernandez, K. Langenfeld, K. R. Wigginton, and A. B. Boehm. 2021. SARS-CoV-2 RNA in wastewater settled solids is associated with COVID-19 cases in a large urban sewershed. *Environmental Science & Technology* 55(1):488–498. https://doi.org/10.1021/acs.est.0c06191.

Green, A. J. E., and G. Zanusso. 2018. Chapter 19: Prion protein amplification techniques. *Handbook of Clinical Neurology* 153:357–370. https://doi.org/10.1016/B978-0-444-63945-5.00019-2.

Grubaugh, N. D., K. Gangavarapu, J. Quick, N. L. Matteson, J. G. De Jesus, B. J. Main, A. L. Tan, L. M. Paul, D. E. Brackney, S. Grewal, N. Gurfield, K. K. A. Van Rompay, S. Isern, S. F. Michael, L. L. Coffey, N. J. Loman, and K. G. Andersen. 2019. An amplicon-based sequencing framework for accurately measuring intrahost virus diversity using PrimalSeq and iVar. *Genome Biology* 20(1):8. https://doi.org/10.1186/s13059-018-1618-7.

Haley, N. J., C. K. Mathiason, S. Carver, M. Zabel, G. C. Telling, and E. A. Hoover. 2011. Detection of chronic wasting disease prions in salivary, urinary, and intestinal tissues of deer: Potential mechanisms of prion shedding and transmission. *Journal of Virology* 85(13):6309–6318. https://doi.org/10.1128/JVI.00425-11.

Hall, W., J. Prichard, P. Kirkbride, R. Bruno, P. K. Thai, C. Gartner, F. Y. Lai, C. Ort, and J. F. Mueller. 2012. An analysis of ethical issues in using wastewater analysis to monitor illicit drug use. *Addiction* 107(10):1767–1773. https://doi.org/10.1111/j.1360-0443.2012.03887.x.

Helfferich, J., M. Knoester, C. C. Van Leer-Buter, R. F. Neuteboom, L. C. Meiners, H. G. Niesters, and O. F. Brouwer. 2019. Acute flaccid myelitis and enterovirus D68: Lessons from the past and present. *European Journal of Pediatrics* 178(9):1305–1315. https://doi.org/10.1007/s00431-019-03435-3.

Hendriksen, R. S., P. Munk, P. Njage, B. van Bunnik, L. McNally, O. Lukjancenko, T. Röder, D. Nieuwenhuijse, S. Karlsmose Pedersen, J. Kjeldgaard, R. S. Kaas, P. T. L. C. Clausen, J. K. Vogt, P. Leekitcharoenphon, M. G. M. van de Schans, T. Zuidema, A. M. de Roda Husman, S. Rasmussen, B. Petersen, The Global Sewage Surveillance project Consortium, C. Amid, G. Cochrane, T. Sicheritz-Ponten, H. Schmitt, J. R. M. Alvarez, A. Aidara-Kane, S. J. Pamp, O. Lund, T. Hald, M. Woolhouse, M. P. Koopmans, H. Vigre, T. Nordahl Petersen, and F. M. Aarestrup. 2019. Global monitoring of antimicrobial resistance based on metagenomics analyses of urban sewage. *Nature Communications* 10:1124. https://doi.org/10.1038/s41467-019-08853-3.

Hoar, C., J. McClary-Gutierrez, M. K. Wolfe, A. Bivins, K. Bibby, A. I. Silverman, and S. L. McLellan. 2022. Looking forward: The role of academic researchers in building sustainable wastewater surveillance programs. *Environmental Health Perspectives* 130(12). https://doi.org/10.1289/EHP11519.

Holm, R. H., J. M. Brick, A. R. Amraotkar, J. L. Hart, A. Mukherjee, J. Zeigler, A. M. Bushau-Sprinkle, L. B. Anderson, K. L. Walker, D. Talley, R. J. Keith, S. N. Rai, K. E. Palmer, A. Bhatnagar, and T. Smith. 2022a. Public awareness of and support for the use of wastewater for SARS-CoV-2 monitoring: A community survey in Louisville, Kentucky. *Environmental Science & Technology: Water*. https://doi.org/10.1021/acsestwater.1c00405.

Holm, R. H., A. Mukherjee, J. P. Rai, R. A. Yeager, D. Talley, S. N. Rai, A. Bhatnagar, and T. Smith. 2022b. SARS-CoV-2 RNA abundance in wastewater as a function of distinct urban sewershed size. *Environmental Science: Water Research & Technology* 8(4):807–819. https://doi.org/10.1039/D1EW00672J.

Hopkins, L., D. Persse, K. Caton, K. Ensor, R. Schneider, C. McCall, and L. B. Stadler. 2022. Citywide wastewater SARS-CoV-2 levels strongly correlated with multiple disease surveillance indicators and outcomes over three COVID-19 waves. *Science of the Total Environment* 855(2023):158967. http://dx.doi.org/10.1016/j.scitotenv.2022.158967.

Horstmann, D. M., J. Emmons, L. Gimpel, T. Subrahmanyan, and J. T. Riordan. 1973. Enterovirus surveillance following a community-wide oral poliovirus vaccination program: A seven-year study. *American Journal of Epidemiology* 97(3):173–186. https://doi.org/10.1093/oxfordjournals.aje.a121498.

Hovi, T., M. Stenvik, H. Partanen, and A. Kangas. 2001. Poliovirus surveillance by examining sewage specimens. Quantitative recovery of virus after introduction into sewerage at remote upstream location. *Epidemiology and Infection* 127(1):101–106. https://doi.org/10.1017/S0950268801005787.

Hrudey, S. E., D. S. Silva, J. Shelley, W. Pons, J. Isaac-Renton, A. H. S. Chik, and B. Conant. 2021. Ethics guidance for environmental scientists engaged in surveillance of wastewater for SARS-CoV-2. *Environmental Science and Technology* 55(13):8484–8491. https://doi.org/10.1021/acs.est.1c00308.

Indiana ACIR (Advisory Committee for Intergovernmental Relations). 2019. More than 11,000 wastewater failures reported in Indiana's unsewered communities.19-C08. https://www.in.gov/iurc/files/IN-Advisory-Commission-March-2019-Indianas-Unsewered-Communities.pdf.

Islam, G., A. Gedge, L. Lara-Jacobo, A. Kirkwood, D. Simmons, and J. P. Desaulniers. 2022. Pasteurization, storage conditions and viral concentration methods influence RT-qPCR detection of SARS-CoV-2 RNA in wastewater. *Science of the Total Environment* 821:153228. https://doi.org/10.1016/j.scitotenv.2022.153228.

Jacobs, D., T. McDaniel, A. Varsani, R. U. Halden, S. Forrest, and H. Lee. 2021. Wastewater monitoring raises privacy and ethical considerations. *IEEE Transactions on Technology and Society* 2(3):116–121. https://doi.org/10.1109/TTS.2021.3073886.

Joh, E. E. 2021. COVID sewage testing as a police surveillance infrastructure. *Notre Dame Journal of Emerging Technologies* 2(2). https://ndlsjet.com/covid-19-sewage-testing-as-a-police-surveillance-infrastructure/.

Kaiser, J. 2020. Poop tests stop COVID-19 outbreak at University of Arizona. *Science News*, August 28. https://www.science.org/content/article/poop-tests-stop-covid-19-outbreak-university-arizona.

Kantor, R. 2022. SARS-CoV-2 variant tracking from wastewater. In *Public Health and Water Conference Proceedings 2022*. Water Environment Federation. https://www.accesswater.org.

Karthikeyan, S., J. I. Levy, P. De Hoff, G. Humphrey, A. Birmingham, K. Jepsen, S. Farmer, H. M. Tubb, T. Valles, C. E. Tribelhorn, R. Tsai, S. Aigner, S. Sathe, N. Moshiri, B. Henson, A. M. Mark, A. Hakim, N. A. Baer, T. Barber, P. Belda-Ferre, M. Chacon, W. Cheung, E. S. Cresini, E. R. Eisner, A. L. Lastrella, E. S. Lawrence, C. A. Marotz, T. T. Ngo, T. Ostrander, A. Plascencia, R. A. Salido, P. Seaver, E. W. Smoot, D. McDonald, R. M. Neuhard, A. L. Scioscia, A. M. Satterlund, E. H. Simmons, D. B. Abelman, D. Brenner, J. C. Bruner, A. Buckley, M. Ellison, J. Gattas, S. L. Gonias, M. Hale, F. Hawkins, L. Ikeda, H. Jhaveri, T. Johnson, V. Kellen, B. Kremer, G. Matthews, R. W. McLawhon, P. Ouillet, D. Park, A. Pradenas, S. Reed, L. Riggs, A. Sanders, B. Sollenberger, A. Song, B. White, T. Winbush, C. M. Aceves, C. Anderson, K. Gangavarapu, E. Hufbauer, E. Kurzban, J. Lee, N. L. Matteson, E. Parker, S. A. Perkins, K. S. Ramesh, R. Robles-Sikisaka, M. A. Schwab, E. Spencer, S. Wohl, L. Nicholson, I. H. McHardy, D. P. Dimmock, C. A. Hobbs, O. Bakhtar, A. Harding, A. Mendoza, A. Bolze, D. Becker, E. T. Cirulli, M. Isaksson, K. M. Schiabor Barrett, N. L. Washington, J. D. Malone, A. M. Schafer, N. Gurfield, S. Stous, R. Fielding-Miller, R. S. Garfein, T. Gaines, C. Anderson, N. K. Martin, R. Schooley, B. Austin, D. R. MacCannell, S. F. Kingsmore, W. Lee, S. Shah, E. McDonald, A. T. Yu, M. Zeller, K. M. Fisch, C. Longhurst, P. Maysent, D. Pride, P. K. Khosla, L. C. Laurent, G. W. Yeo, K. G. Andersen, and R. Knight. 2022. Wastewater sequencing reveals early cryptic SARS-CoV-2 variant transmission. *Nature* 609:101–108. https://doi.org/10.1038/s41586-022-05049-6.

Karthikeyan, S., A. Nguyen, D. McDonald, Y. Zong, N. Ronquillo, J. Ren, J. Zou, S. Farmer, G. Humphrey, D. Henderson, T. Javidi, K. Messer, C. Anderson, R. Schooley, N. K. Martin, and R. Knight. 2021. Rapid, large-scale wastewater surveillance and automated reporting system enable early detection of nearly 85% of COVID-19 cases on a university campus. *mSystems* 6(4):e0079321. https://doi.org/10.1128/mSystems.00793-21.

Keshaviah, A., I. Huff, X. C. Hu, V. Guidry, A. Christensen, S. Berkowitz, S. Reckling, S. McLellan, A. Roguet, and I. Mussa. 2022a. Separating signal from noise in wastewater data: An algorithm to identify community-level COVID-19 surges. *medRxiv*. https://doi.org/10.1101/2022.09.19.22280095.

Keshaviah, A., R. N. Karmali, D. Vohra, T. Huffman, X. C. Hu, and M. B. Diamond. 2022b. *The role of wastewater data in pandemic management*. Washington, DC: Mathematica. https://www.rockefellerfoundation.org/wp-content/uploads/2022/04/The-Role-of-Wastewater-Data-in-Pandemic-Management-Survey-Research-Brief-Final.pdf.

Kim, H., R. G. Webster, and R. J. Webby. 2018. Influenza virus: Dealing with a drifting and shifting pathogen. *Viral Immunology* 31(2):174–183. https://doi.org/10.1089/vim.2017.0141.

Kirby, A. E. 2022. National wastewater surveillance system: Implementation overview. Presentation at Meeting 1 of the NASEM Committee on Community Wastewater-based Infectious Disease Surveillance. Virtual meeting. April 21. https://www.nationalacademies.org/event/04-21-2022/community-wastewater-based-infectious-disease-surveillance-meeting-1-public-session.

Kirby, A. E., M. S. Walters, W. C. Jennings, R. Fugitt, N. LaCross, M. Mattioli, Z. A. Marsh, V. A. Roberts, J. W. Mercante, J. Yoder, and V. R. Hill. 2021. Using wastewater surveillance data to support the COVID-19 response—United States, 2020–2021. *Morbidity and Mortality Weekly Report* 70(36):1242–1244. http://dx.doi.org/10.15585/mmwr.mm7036a2.

Kirby, A. E., R. M. Welsh, Z. A. Marsh, A. T. Yu, D. J. Vugia, A. B. Boehm, M. K. Wolfe, B. J. White, S. R. Matzinger, A. Wheeler, L. Bankers, K. Andresen, C. Salatas, New York City Department of Environmental Protection, D. A. Gregory, M. C. Johnson, M. Trujillo, S. Kannoly, D. S. Smyth, J. J. Dennehy, N. Sapoval, K. Ensor, T. Treangen, L. B. Stadler, and L. Hopkins. 2022. Notes from the field: Early Evidence of the SARS-CoV-2 B.1.1.529 (Omicron) variant in community wastewater–United States, November–December 2021. *Morbidity and Mortality Weekly Report* 71(3):103–105. https://doi.org/10.15585%2Fmmwr.mm7103a5.

Klapsa, D., T. Wilton, A. Zealand, E. Bujaki, E. Saxentoff, C. Troman, A. G. Shaw, A. Tedcastle, M. Majumdar, R. Mate, J. O. Akello, S. Huseynov, A. Zeb, M. Zambon, A. Bell, J. Hagan, M. J. Wade, M. Ramsay, N. C. Grassly, V. Saliba, and J. Martin. 2022. Sustained detection of type 2 poliovirus in London sewage between February and July, 2022, by enhanced environmental surveillance. *The Lancet* 400(10362):1531–1538. https://doi.org/ 10.1016/S0140-6736(22)01804-9.

Kline, K. E., J. Shover, A. J. Kallen, D. R. Lonsway, S. Watkins, and J. R. Miller. 2016. Investigation of first identified *mcr-1* gene in an isolate from a U.S. patient—Pennsylvania, 2016. *Morbidity and Mortality Weekly Report* 65(36):977–978. http://dx.doi.org/10.15585/mmwr.mm6536e2.

Klingler, C., D. S. Silva, C. Schuermann, A. A. Reis, A. Saxena, and D. Strech. 2017. Ethical issues in public health surveillance: A systematic qualitative review. *BMC Public Health* 17(295). https://doi.org/10.1186/s12889-017-4200-4.

Krüger, D., A. Thomzig, G. Lenz, K. Kampf, P. McBride, and M. Beekes. 2009. Faecal shedding, alimentary clearance and intestinal spread of prions in hamsters fed with scrapie. *Veterinary Research* 40(4). https://doi.org/10.1051/vetres:2008042.

LaJoie, A. S., R. H. Holm, L. B. Anderson, H. D. Ness, and T. Smith. 2022. Survey of nationwide public perceptions regarding acceptance of wastewater used for community health monitoring in the United States. *medRxiv*. https://doi.org/10.1101/2022.03.16.22272262.

Layton, B. A., D. Kaya, C. Kelly, K. J. Williamson, D. Alegre, S. M. Bachhuber, P. G. Banwarth, J. W. Bethel, K. Carter, B. D. Dalziel, M. Dasenko, M. Geniza, A. George, A. M. Girard, R. Haggerty, K. A. Higley, D. M. Hynes, J. Lubchenco, K. R. McLaughlin, F. J. Nieto, A. Noakes, M. Peterson, A. D. Piemonti, J. L. Sanders, B. M. Tyler, and T. S. Radniecki. 2022. Evaluation of a wastewater-based epidemiological approach to estimate the prevalence of SARS-CoV-2 infections and the detection of viral variants in disparate Oregon communities at city and neighborhood scales. *Environmental Health Perspectives* 130(6):67010. https://ehp.niehs.nih.gov/doi/10.1289/EHP10289.

Lee, L. M., C. M. Heilig, and A. White. 2012. Ethical justification for conducting public health surveillance without patient consent. *American Journal of Public Health* 102:38–44. https://doi.org/10.2105/AJPH.2011.300297.

Lee, W. L., X. Gu, F. Armas, F. Chandra, H. Chen, F. Qu, M. Leifels, A. Xiao, F. J. D. Chua, G. W. C. Kwok, S. Jolly, C. Y. J. Lim, J. Thompson, and E. J. Alm. 2021. Quantitative SARS-CoV-2 tracking of variants Delta, Delta plus, Kappa and Beta in wastewater by allele-specific RT-qPCR. *medRxiv*. https://doi.org/10.1101/2021.08.03.21261298.

Li, B., D. Yoong Wen Di, P. Saingam, M. K. Jeon, and T. Yan. 2021. Fine-scale temporal dynamics of SARS-CoV-2 RNA abundance in wastewater during a COVID-19 lockdown. *Water Research* 197:117093. https://doi.org/10.1016/j.watres.2021.117093.

Link-Gelles, R., E. Lutterloh, P. Schnabel Ruppert, P. B. Backenson, K. St George, E. S. Rosenberg, B. J. Anderson, M. Fuschino, M. Popowich, C. Punjabi, M. Souto, K. McKay, S. Rulli, T. Insaf, D. Hill, J. Kumar, I. Gelman, J. Jorba, T. F. F. Ng, N. Gerloff, N. B. Masters, A. Lopez, K. Dooling, S. Stokley, S. Kidd, M. S. Oberste, J. Routh, and 2022 U.S. Poliovirus Response Team. 2022. Public health response to a case of paralytic poliomyelitis in an unvaccinated person and detection of poliovirus in wastewater—New York, June–August 2022. *Morbidity and Mortality Weekly Report* 71(33):1065–1068. https://doi.org/10.15585/mmwr.mm7133e2.

Liu, Y. Y., Y. Wang, T. R. Walsh, L. X. Yi, R. Zhang, J. Spencer, Y. Doi, G. Tian, B. Dong, X. Huang, L. F. Yu, D. Gu, H. Ren, X. Chen, L. Lv, D. He, H. Zhou, Z. Liang, J. H. Liu, and J. Shen. 2016. Emergence of plasmid-mediated colistin resistance mechanism MCR-1 in animals and human beings in China: A microbiological and molecular biological study. *The Lancet* 16(2):161–168. https://doi.org/10.1016/S1473-3099(15)00424-7.

Long, A. S., A. L. Hanlon, and K. L. Pellegrin. 2018. Socioeconomic variables explain rural disparities in US mortality rates: Implications for rural health research and policy. *SSM-Population Health* 6:72–74.

Lu, F. S., M. W. Hattab, C. L. Clemente, M. Biggerstaff, and M. Santillana. 2019. Improved state-level influenza nowcasting in the United States leveraging Internet-based data and network approaches. *Nature Communications* 10(1):1–10.

Majumdar, M., T. Wilton, Y. Hajarha, D. Klapsa, and J. Martin. 2019. Detection of enterovirus D68 in wastewater samples from the United Kingdom during outbreaks reported globally between 2015 and 2018. *BioRxiv*. https://doi.org/10.1101/738948.

Maksimovic Carvalho Ferreira, O., Ž. Lengar, Z. Kogej, K. Ba nik, I. Bajde, M. Milavec, A. Županič, N. Mehle, D. Kutnjak, M. Ravnikar, and I. Gutierrez-Aguirre. 2022. Evaluation of methods and processes for robust monitoring of SARS-CoV-2 in wastewater. *Food and Environmental Virology* 1–17. https://doi.org/10.1007/s12560-022-09533-0.

Manning, S., and M. Walton. 2021. *COVID-19 surveillance in wastewater: Communications and equity*. New Zealand Crown Research Institute of Environmental Science and Research. https://www.esr.cri.nz/assets/ESR-2021-Covid-in-Wastewater-social-summary.pdf.

Manor, Y., R. Handsher, T. Halmut, M. Neuman, B. Abramovitz, A. Mates, and E. Mendelson. 1999a. A double-selective tissue culture system for isolation of wild-type poliovirus from sewage applied in a long-term environmental surveillance. *Applied and Environmental Microbiology* 65(4):1794–1797. https://doi.org/10.1128/AEM.65.4.1794-1797.1999.

Manor, Y., R. Handsher, T. Halmut, M. Neuman, A. Bobrov, H. Rudich, A. Vonsover, L. Shulman, O. Kew, and E. Mendelson. 1999b. Detection of poliovirus circulation by environmental surveillance in the absence of clinical cases in Israel and the Palestinian Authority. *Journal of Clinical Microbiology* 37(6):1670–1675. https://doi.org/10.1128/JCM.37.6.1670-1675.1999.

Marques, E., E. E. Da Silva, V. M. Dos Santos, O. M. Kew, and M. T. Martins. 1993. Application of the polymerase chain reaction (PCR) to poliomyelitis surveillance through the analyses of sewage samples. *World Journal of Microbiology and Biotechnology* 9(5):566–569. https://doi.org/10.1007/BF00386295.

Mataraci-Kara, E., M. Ataman, G. Yilmaz, and B. Ozbek-Celik. 2020. Evaluation of antifungal and disinfectant-resistant Candida species isolated from hospital wastewater. *Archives of Microbiology* 202(9):2543–2550. https://doi.org/10.1007/s00203-020-01975-z.

McClary-Gutierrez, J. S., M. C. Mattioli, P. Marcenac, A. I. Silverman, A. B. Boehm, K. Bibby, M. Balliet, F. L. de los Reyes, D. Gerrity, J. F. Griffith, P. A. Holden, D. Katehis, G. Kester, N. LaCross, E. K. Lipp, J. Meiman, R. T. Noble, D. Brossard, and S. L. McLellan. 2021. SARS-CoV-2 wastewater surveillance for public health action. *Emerging Infectious Diseases* 27(9):1–8. https://doi.org/10.3201/eid2709.210753.

McMahan, C. S., S. Self, L. Rennert, C. Kalbaugh, D. Kriebel, D. Graves, C. Colby, J. A. Deaver, S. C. Popat, T. Karanfil, and D. L. Freedman. 2021. COVID-19 wastewater epidemiology: A model to estimate infected populations. *The Lancet Planetary Health* 5(12):e874–e881. https://doi.org/10.1016/S2542-5196(21)00230-8.

Medema, G., L. Heijnen, G. Elsinga, R. Italiaander, and A. Brouwer. 2020. Presence of SARS-Coronavirus-2 RNA in sewage and correlation with reported COVID-19 prevalence in the early stage of the epidemic in The Netherlands. *Environmental Science & Technology Letters* 7(7):511–516. https://doi.org/10.1021/acs.estlett.0c00357.

Medina, C. Y., K. Kadonsky, F. A. R. Roman, A. Q. Tariqi, R. Sinclair, P. M. D'Aoust, R. Delatolla, H. Bischel, and C. C. Naughton. 2022. The need of an environmental justice approach for wastewater based epidemiology for rural and disadvantaged communities: A review in California. *Current Opinion in Environmental Science & Health* 27:100348. https://doi.org/10.1016%2Fj.coesh.2022.100348.

Melnick, J. L. 1947. Poliomyelitis virus in urban sewage in epidemic and in nonepidemic times. *American Journal of Epidemiology* 45(2):240–253. https://doi.org/10.1093/oxfordjournals.aje.a119132.

Mendoza Grijalva, L., B. Brown, A. Cauble, and W. A. Tarpeh. 2022. Diurnal variability of SARS-CoV-2 RNA concentrations in hourly grab samples of wastewater influent during low COVID-19 incidence. *Environmental Science and Technology: Water*. https://doi.org/10.1021/acsestwater.2c00061.

Mercier, E., P. M. D'Aoust, O. Thakali, N. Hegazy, J. J. Jia, Z. Zhang, W. Eid, J. Plaza-Diaz, M. P. Kabir, W. Fang, A. Cowan, S. E. Stephenson, L. Pisharody, A. E. MacKenzie, T. E. Graber, S. Wan, and R. Delatolla. 2022. Wastewater surveillance of influenza activity: Early detection, surveillance, and subtyping in city and neighbourhood communities. *medRxiv*. https://doi.org/10.1101/2022.06.28.22276884.

Miller, M. W., and E. S. Williams. 2004. Chronic wasting disease of cervids. *Current Topics in Microbiology and Immunology* 284:193–214. https://doi.org/10.1007/978-3-662-08441-0_8.

NASEM (National Academies of Sciences, Engineering, and Medicine). 2022. *Combating antimicrobial resistance and protecting the miracle of modern medicine*. Washington, DC: The National Academies Press. https://doi.org/10.17226/26350.

Natarajan, A., S. Zlitni, E. F. B. Brooks, S. E. Vance, A. Dahlen, H. Hedlin, R. M. Park, A. Han, D. T. Schmidtke, R. Verma, K. B. Jacobson, J. Parsonnet, H. F. Bonilla, U. Singh, B. A. Pinsky, J. R. Andrews, P. Jagannathan, and A. S. Bhatt. 2022. Gastrointestinal symptoms and fecal shedding of SARS-CoV-2 RNA suggest prolonged gastrointestinal infection. *Med (New York, New York)* 3(6):371–387.e9.https://doi.org/10.1016/j.medj.2022.04.001.

Naughton, C. C., F. A. Roman Jr., A. G. F. Alvarado, A. Q. Tariqi, M. A. Deeming, K. Bibby, A. Bivins, J. B. Rose, G. Medema, W. Ahmed, P. Katsivelis, V. Allan, R. Sinclair, Y. Zhang, and M. N. Kinyua. 2021. Show us the data: Global COVID-19 wastewater monitoring efforts, equity, and gaps. *medRxiv* https://doi.org/10.1101/2021.03.14.21253564.

NDWAC (National Drinking Water Advisory Council). 2004. *National Drinking Water Advisory Council report on the CCL Classification Process to the U.S. Environmental Protection Agency.* https://www.epa.gov/sites/default/files/2015-11/documents/report_ccl_ndwac_07-06-04.pdf.

Nelson, B. 2022. What poo tells us: Wastewater surveillance comes of age amid covid, monkeypox, and polio. *BMJ* 378. https://doi.org/10.1136/bmj.o1869.

Nelson, W. W., T. A. Scott, M. Boules, C. Teigland, A. Parente, S. Unni, and P. Feuerstadt. 2021. Health care resource utilization and costs of recurrent *Clostridioides difficile* infection in the elderly: A real-world claims analysis. *Journal of Managed Care and Specialty Pharmacy* 27(7):828–838. https://doi.org/10.18553/jmcp.2021.20395.

Nicholson, E. M. 2015. Detection of the disease-associated form of the prion protein in biological samples. *Future Science* 7(2). https://doi.org/10.4155/bio.14.301.

Nordahl Petersen, T., S. Rasmussen, H. Hasan, C. Carøe, J. Bælum, A. C. Schultz, L. Bergmark, C. A. Svendsen, O. Lund, T. Sicheritz-Pontén, and F. M. Aarestrup. 2015. Meta-genomic analysis of toilet waste from long distance flights: A step towards global surveillance of infectious diseases and antimicrobial resistance. *Scientific Reports* 5:11444. https://doi.org/10.1038/srep11444.

NRC (National Research Council). 2001. *Classifying drinking water contaminants for regulatory consideration.* Washington, DC: National Academy Press. https://doi.org/10.17226/10080.

NSF (National Science Foundation). 2009. Chapter II - Proposal Preparation Instructions. https://www.nsf.gov/pubs/policydocs/pappguide/nsf09_1/gpg_2.jsp.

Parnanen, K. M. M., C. Narciso-da-Rocha, D. Kneis, T. U. Beredonk, D. Cacace, T. T. Do, C. Elpers, D. Fatta-Kassinos, I. Henriques, T. Jaeger, A. Karkman, J. L. Martinez, S. G. Michael, I. Michael-Kordatou, K. O'Sullivan, S. Rodriguez-Mozaz, T. Schwartz, H. Sheng, H. Sørum, R. D. Stedtfeld, J. M. Tiedje, S. Varella della Giustina, F. Walsh, I. Vaz-Moreira, and C. M. Manaia. 2019. Antibiotic resistance in European wastewater treatment plants mirrors the pattern of clinical antibiotic resistance prevalence. *Science Advances* 5(3). https://doi.org/10.1126/sciadv.aau9124.

Paul, J. R., and J. D. Trask. 1941. The virus of poliomyelitis in stools and sewage. *Journal of the American Medical Association* 116(6):493–498. https://doi.org/10.1001/jama.1941.02820060041009.

Paul, J. R., J. D. Trask, and S. Gard. 1940. II. Poliomyelitic virus in urban sewage. *The Journal of Experimental Medicine* 71(6):765–777. https://doi.org/10.1084/jem.71.6.765.

Peccia, J., A. Zulli, D. E. Brackney, N. D. Grubaugh, E. H. Kaplan, A. Casanovas-Massana, A. I. Ko, A. A. Malik, D. Wang, M. Wang, J. L. Warren, D. M. Weinberger, W. Arnold, and S. B. Omer. 2020. Measurement of SARS-CoV-2 RNA in wastewater tracks community infection dynamics. *Nature Biotechnology* 38(10):1164–1167. https://doi.org/10.1038/s41587-020-0684-z.

Pecson, B. M, E. Darby, C. Haas, Y. Amha, M. Bartolo, R. Danielson, Y. Dearborn, G. Di Giovanni, C. Ferguson, S. Fevig, E. Gaddis, D. Gray, G. Lukasik, B. Mull, A. Olivieri, Y. Qu, and SARS-CoV-2 Interlaboratory Consortium. 2021. Reproducibility and sensitivity of 36 methods to quantify the SARS-CoV-2 genetic signal in raw wastewater: Findings from an interlaboratory methods evaluation in the U.S. *Environmental Science: Water Research & Technology* 7:504–520. https://doi.org/10.1039/D0EW00946F.

Philo, S. E., A. Q. W. Ong, E. K. Keim, R. Swanstrom, A. L. Kossik, N. A. Zhou, N. K. Beck, and J. S. Meschke. 2022. Development and validation of the skimmed milk pellet extraction protocol for SARS-CoV-2 wastewater surveillance. *Food and Environmental Virology*. https://doi.org/10.1007/s12560-022-09512-5.

Polo, D., M. Quintela-Baluja, A. Corbishley, D. L. Jones, A. C. Singer, D. W. Graham, and J. L. Romalde. 2020. Making waves: Wastewater-based epidemiology for COVID-19–approaches and challenges for surveillance and prediction. *Water Research* 186:116404. https://doi.org/10.1016/j.watres.2020.116404.

Pöyry, T., M. Stenvik, and T. Hovi. 1988. Viruses in sewage waters during and after a poliomyelitis outbreak and subsequent nationwide oral poliovirus vaccination campaign in Finland. *Applied and Environmental Microbiology* 54(2):371–374. https://doi.org/10.1128/aem.54.2.371-374.1988.

Rader, B., A. Gertz, A. D. Iuliano, M. Gilmer, L. Wronski, C. M. Astley, K. Sewalk, T. J. Varrelman, J. Cohen, R. Parikh, H. E. Reese, C. Reed, and J. S. Brownstein. 2022. Use of at-home COVID-19 tests—United States, August 23, 2021–March 12, 2022. *Morbidity and Mortality Weekly Report* 71(13):489–494. http://dx.doi.org/10.15585/mmwr.mm7113e1.

Ram, N., L. Gable, and J. L. Ram. 2022. The future of wastewater monitoring for the public health. *University of Richmond Law Review* 56(911):911–952. https://lawreview.richmond.edu/files/2022/05/Ram-563-online.pdf.

Ranta, J., T. Hovi, and E. Arjas. 2001. Poliovirus surveillance by examining sewage water specimens: Studies on detection probability using simulation models. *Risk Analysis* 21(6):1087–1096. https://doi.org/10.1111/0272-4332.t01-1-216174.

Ritchey, M. D., H. G. Rosenblum, K. Del Guercio, M. Humbard, S. Santos, J. Hall, J. Chaitram, and R. M. Salerno. 2022. COVID-19 self-test data: Challenges and opportunities—United States, October 31, 2021–June 11, 2022. *Morbidity and Mortality Weekly Report* 71:1005–1010. http://dx.doi.org/10.15585/mmwr.mm7132a1.

Rolfes, M. A., I. M. Foppa, S. Garg, B. Flannery, L. Brammer, J. A. Singleton, E. Burns, D. Jernigan, S. J. Olsen, J. Bresee, and C. Reed. 2018. Annual estimates of the burden of seasonal influenza in the United States: A tool for strengthening influenza surveillance and preparedness. *Influenza and Other Respiratory Viruses* 12(1):132–137. https://doi.org/10.1111/irv.12486.

Ryerson, A. B., D. Lang, M. A. Alazawi, M. Neyra, D. T. Hill, K. St George, M. Fuschino, E. Lutterloh, B. Backenson, S. Rulli, P. S. Ruppert, J. Lawler, N. McGraw, A. Knecht, I. Gelman, J. R. Zucker, E. Omoregie, S. Kidd, D. E. Sugerman, J. Jorba, Gerloff, T. F. F. Ng, A. Lopez, N. B. Masters, J. Leung, C. C. Burns, J. Routh, S. R. Bialek, M. S. Oberste, and E. S. Rosenberg. 2022. Wastewater testing and detection of poliovirus type 2 genetically linked to virus isolated from a paralytic polio case - New York, March 9-October 11, 2022. 2022. *Morbidity and Mortality Weekly Report* 71(44):1418–1424. https://doi.org/10.15585/mmwr.mm7144e2.

Saad-Roy, C. M., C. J. E. Metcalf, and B. T. Grenfell. 2022. Immuno-epidemiology and the predictability of viral evolution. *Science* 376(6598):1161–1162. https://doi.org/10.1126/science.abn9410.

Safar, J. G., P. Lessard, G. Tamgüney, Y. Freyman, C. Deering, F. Letessier, S. J. Dearmond, and S. B. Prusiner. 2008. Transmission and detection of prions in feces. *Journal of Infectious Diseases* 198(1):81–89. https://doi.org/10.1086/588193.

Safford, H., K. Shapiro, and H. N. Bischell. 2022. Wastewater analysis can be a powerful public health tool—if it's done sensibly. *Proceedings of the National Academy of Sciences* 119(6):e2119600119. https://www.pnas.org/doi/10.1073/pnas.2119600119.

Saguti, F., E. Magnil, L. Enache, M. P. Churqui, A. Johansson, D. Lumley, F. Davidsson, L. Dotevall, A. Mattsson, E. Trybala, and M. Lagging. 2021. Surveillance of wastewater revealed peaks of SARS-CoV-2 preceding those of hospitalized patients with COVID-19. *Water Research* 189:116620. https://doi.org/10.1016/j.watres.2020.116620.

Scassa, T., P. J. Robinson, and R. Mosoff. 2022. The datafication of wastewater: Legal, ethical and civic considerations. *Technology and Regulation*: 23–35. https://doi.org/10.26116/techreg.2022.003.

Schussman, M. K., A. Roguet, A. Schmoldt, B. Dinan, and S. L. McLellan. 2022. Wastewater surveillance using ddPCR reveals highly accurate tracking of Omicron variant due to altered N1 probe binding efficiency. *medRxiv*. https://doi.org/10.1101/2022.02.18.22271188.

Sharara, N., N. Endo, C. Duvallet, N. Ghaeli, M. Matus, J. Heussner, S. W. Olesen, E. J. Alm, P. R. Chai, and T. B. Erickson. 2021. Wastewater network infrastructure in public health: Applications and learnings from the COVID-19 pandemic. *PLOS Global Public Health* 1(12). https://doi.org/10.1371/journal.pgph.0000061.

Sims, N., L. Avery, and B. Kasprzyk-Hordern. 2021. *Review of wastewater monitoring applications for public health and novel aspects of environmental quality (CD2020_07)*. Scotland's Centre of Expertise for Waters. https://www.crew.ac.uk/sites/www.crew.ac.uk/files/publication/FINAL_REPORT_FOR_Review%20of%20wastewater%20monitoring%20applications.pdf.

Smyth, D. S., M. Trujillo, D. A. Gregory, K. Cheung, A. Gao, M. Graham, Y. Guan, C. Guldenpfennig, I. Hoxie, S. Kannoly, N. Kubota, T. D. Lyddon, M. Markman, C. Rushford, K. M. San, G. Sompanya, F. Spagnolo, R. Suarez, E. Teixeiro, M. Daniels, M. C. Johnson, and J. J. Dennehy. 2022. Tracking cryptic SARS-CoV-2 lineages detected in NYC wastewater. *Nature Communications* 13(1):1836. https://doi.org/10.1038/s41467-022-29573-1.

Snyder, A., and S. Cullinane. 2020. Some scientists are using sewage to measure the prevalence of coronavirus in their communities. *CNN News*, April 26. https://www.cnn.com/2020/04/26/us/covid-19-sewage-testing/index.html.

Soller, J., W. Jennings, M. Schoen, A. Boehm, K. Wigginton, R. Gonzalez, K. E. Graham, G. McBride, A. Kirby, and M. Mattioli. 2022. Modeling infection from SARS-CoV-2 wastewater concentrations: Promise, limitations, and future directions. *Journal of Water and Health* 20(8):1197–1211. https://doi.org/10.2166/wh.2022.094.

Sooksawasdi Na Ayudhya, S., B. M. Laksono, and D. van Riel. 2021. The pathogenesis and virulence of enterovirus-D68 infection. *Virulence* 12(1):2060–2072. https://doi.org/10.1080/21505594.2021.1960106.

Sousan, S., M. Fan, K. Outlaw, S. Williams, and R. L. Roper. 2022. SARS-CoV-2 detection in air samples from inside heating, ventilation, and air conditioning (HVAC) systems-COVID surveillance in student dorms. *American Journal of Infection Control* 50(3):330–335. https://doi.org/10.1016/j.ajic.2021.10.009/.

Stadler, L. B., K. B. Ensor, J. R. Clark, P. Kalvapalle, Z. W. LaTurner, L. Mojica, A. Terwilliger, Y. Zhuo, P. Ali, V. Avadhanula, R. Bertolusso, T. Crosby, H. Hernandez, M. Hollstein, K. Weesner, D. M. Zong, D. Persse, P. A. Piedra, A. W. Maresso, and L. Hopkins. 2020. Wastewater analysis of SARS-CoV-2 as a predictive metric of positivity rate for a major metropolis. *medRxiv*. https://doi.org/10.1101/2020.11.04.20226191.

Tamburini, F. B., T. M. Andermann, E. Tkachenko, F. Senchyna, N. Banaei, and A. S. Bhatt. 2018. Precision identification of diverse bloodstream pathogens in the gut microbiome. *Nature Medicine* 24:1809–1814. https://doi.org/10.1038/s41591-018-0202-8.

Tchobanoglous, G., F. L. Burton, and H. D. Stensel. 2003. *Wastewater engineering: Treatment and reuse*. 4th ed. New York: McGraw-Hill.

Tedcastle, A., T. Wilton, E. Pegg, D. Klapsa, E. Bujaki, R. Mate, M. Fritzsche, M. Majumdar, and J. Martin. 2022. Detection of enterovirus D68 in wastewater samples from the UK between July and November 2021. *Viruses* 14(1):143. https://doi.org/10.3390/v14010143.

Tennant, J. M., M. Li, D. M. Henderson, M. L. Tyer, N. D. Denkers, N. J. Haley, C. K. Mathiason, and E. A. Hoover. 2020. Shedding and stability of CWD prion seeding activity in cervid feces. *PLOS One* 15(3):e0227094. https://doi.org/10.1371/journal.pone.0227094.

Terry, L. A., L. Howells, K. Bishop, C. A. Baker, S. Everest, L. Thorne, B. C. Maddison, and K. C. Gough. 2011. Detection of prions in the faeces of sheep naturally infected with classical scrapie. *Veterinary Research* 42(65). https://doi.org/10.1186/1297-9716-42-65.

Thompson, W. W., D. K. Shay, E. Weintraub, L. Brammer, C. B. Bridges, N. J. Cox, and K. Fukuda. 2004. Influenza-associated hospitalizations in the United States. *JAMA* 292(11):1333–1340. https://doi.org/10.1001/jama.292.11.1333.

Thraenhart, O., E. Kuwert, and W. Worringen. 1977. Modellversuche zur Uberwachung der poliomyelitisgefährdung der Bevölkerung einer Grossstadt (Essen) im Ruhrgebiet durch Abwasseruntersuchungen [Experiments on poliomyelitis surveillance in an urban population of the Ruhr Valley (Essen) by means of virological investigations of the sewage water (author's transl)]. *Zentralblatt für Bakteriologie, Parasitenkunde, Infektionskrankheiten und Hygiene. Erste Abteilung Originale. Reihe B: Hygiene, Präventive Medizin* 164(4):328–339.

U.S. Census Bureau. 2022. *The 2021 American Housing Survey*. https://www.census.gov/programs-surveys/ahs/data/2021/ahs-2021-public-use-file--puf-/ahs-2021-national-public-use-file--puf-.html.

van der Avoort, H. G., J. H. Reimerink, A. Ras, M. N. Mulders, and A. M. van Loon. 1995. Isolation of epidemic poliovirus from sewage during the 1992-3 type 3 outbreak in The Netherlands. *Epidemiology and Infection* 114(3):481–491. https://doi.org/10.1017/s0950268800052195.

Vogel, G. 2022. Signals from the sewer: Measuring virus levels in wastewater can help track the pandemic. But how useful is that? *Science* 375(6585):1100–1104. https://www.science.org/doi/10.1126/science.adb1874.

Wade, M. J., A. L. Jacomo, E. Armenise, M. R. Brown, J. T. Bunce, G. J. Cameron, Z. Fang, D. F. Gilpin, D. W. Graham, J. M. Grimsley, and A. Hart. 2022. Understanding and managing uncertainty and variability for wastewater monitoring beyond the pandemic: Lessons learned from the United Kingdom national COVID-19 surveillance programmes. *Journal of Hazardous Materials* 424:127456. https://doi.org/10.1016/j.jhazmat.2021.127456.

Ward, I. L., C. Bermingham, D. Ayoubkhani, O. J. Gethings, K. Pouwels, T. Yates, K. Khunti, J. Hippisley-Cox, A. Banerjee, A. S. Walker, and V. Nafilyan. 2022. Risk of COVID-19 related deaths for SARS-CoV-2 Omicron (B.1.1.529) compared with Delta (B.1.617.2). *medRxiv*. https://doi.org/10.1136/bmj-2022-070695.

Ward, T., and A. Johnsen. 2021. Understanding an evolving pandemic: An analysis of the clinical time delay distributions of COVID-19 in the United Kingdom. *PLOS One* 16(10):e0257978. https://doi.org/10.1371/journal.pone.0257978.

Wehrendt, D. P., M. G. Massó, A. Gonzales Machuca, C. V. Vargas, M. E. Barrios, J. Campos, D. Costamagna, L. Bruzzone, D. M. Cisterna, N. G. Iglesias, V. A. Mbayed, E. Baumeister, D. Centrón, M. P. Quiroga, and L. Erijman. 2021. A rapid and simple protocol for concentration of SARS-CoV-2 from sewage. *Journal of Virological Methods* 297:114272. https://doi.org/10.1016/j.jviromet.2021.114272.

Weil, M., M. Mandelboim, E. Mendelson, Y. Manor, L. Shulman, D. Ram, G. Barkai, Y. Shemer, D. Wolf, Z. Kra-Oz, and L. Weiss. 2017. Human enterovirus D68 in clinical and sewage samples in Israel. *Journal of Clinical Virology* 86:52–55. https://doi.org/10.1016/j.jcv.2016.11.013.

WHO (World Health Organization). 2015. *Ebola situation report: 18 November 2015.* https://apps.who.int/iris/bitstream/handle/10665/195839/ebolasitrep_18Nov2015_eng.pdf?sequence=1&isAllowed=y.

WHO. 2017. *WHO guidelines on ethical issues in public health surveillance.* Geneva: World Health Organization. License: CC BY-NC-SA 3.0 IGO. https://apps.who.int/iris/bitstream/handle/10665/255721/9789241512657-eng.pdf.

Wise, J. 2022. Poliovirus is detected in sewage from north and east London. *BMJ* 377:o1546. https://doi.org/10.1136/bmj.o1546.

Wolfe, M. K., D. Duong, K. M. Bakker, M. Ammerman, L. Mortenson, B. Hughes, P. Arts, A. S. Lauring, W. J. Fitzsimmons, E. Bendall, C. E. Hwang, E. T. Martin, B. J. White, A. B. Boehm, and K. R. Wigginton. 2022. Wastewater-based detection of two influenza outbreaks. *Environmental Science and Technology Letters* 9(8):687–692. https://doi.org/10.1021/acs.estlett.2c00350.

Wu, F., A. Xiao, J. Zhang, K. Moniz, N. Endo, F. Armas, R. Bonneau, M. A. Brown, M. Bushman, P. R. Chai, C. Duvallet, T. B. Erickson, K. Foppe, N. Ghaeli, X. Gu, W. P. Hanage, K. H. Huang, W. L. Lee, M. Matus, K. A. McElroy, J. Nagler, S. F. Rhode, M. Santillana, J. A. Tucker, S. Wuertz, S. Zhao, J. Thompson, and E. J. Alm. 2022a. SARS-CoV-2 RNA concentrations in wastewater foreshadow dynamics and clinical presentation of new COVID-19 cases. *Science of the Total Environment* 805:150121. https://doi.org/10.1016/j.scitotenv.2021.150121.

Wu, Y., L. Kang, Z. Guo, J. Liu, M. Liu, and W. Liang. 2022b. Incubation period of COVID-19 caused by unique SARS-CoV-2 strains: A systematic review and meta-analysis. *JAMA Network Open* 5(8):e222800. https://doi.org/10.1001/jamanetworkopen.2022.28008.

Xiao, A., F. Wu, M. Bushman, J. Zhang, M. Imakaev, P. R. Chai, C. Duvallet, N. Endo, T. B. Erickson, F. Armas, B. Arnold, H. Chen, F. Chandra, N. Ghaeli, X. Gu, W. P. Hanage, W. L. Lee, M. Matus, K. A. McElroy, K. Moniz, S. F. Rhode, J. Thompson, and E. J. Alm. 2022. Metrics to relate COVID-19 wastewater data to clinical testing dynamics. *Water Research* 212:118070. https://doi.org/10.1016/j.watres.2022.118070.

Ye, Y., R. M. Ellenberg, K. E. Graham, and K. R. Wigginton. 2016. Survivability, partitioning, and recovery of enveloped viruses in untreated municipal wastewater. *Environmental Science and Technology* 50(10):5077–5085. https://doi.org/10.1021/acs.est.6b00876.

Yeager, R., R. H. Holm, K. Saurabh, J. L. Fuqua, D. Talley, A. Bhatnagar, and T. Smith. 2021. Wastewater sample site selection to estimate geographically resolved community prevalence of COVID-19: A sampling protocol perspective. *GeoHealth* 5(7):e2021GH000420. https://doi.org/10.1029/2021GH000420.

Yokoyama, T. 1999. The immunodetection of the abnormal isoform of prion protein. *The Histochemical Journal* 31(4):209–212. https://doi.org/10.1023/a:1003514021800.

Yong, D., M. A. Toleman, C. G. Giske, H. S. Cho, K. Sundman, K. Lee, and T. R. Walsh. 2009. Characterization of a new metallo-β-lactamase gene, $bla_{NDM-1}$, and a novel erythromycin esterase gene carried on a unique genetic structure in *Klebsiella pneumoniae* sequence Type 14 from India. *Antimicrobial Agents and Chemotherapy* 53(12):5046–5054. https://doi.org/10.1128/AAC.00774-09.

Yu, A. T., B. Hughes, M. K. Wolfe, T. Leon, D. Duong, A. Rabe, L. C. Kennedy, S. Ravuri, B. J. White, K. R. Wigginton, A. B. Boehm, and D. J. Vugia. 2022. Estimating relative abundance of 2 SARS-CoV-2 variants through wastewater surveillance at 2 large metropolitan sites, United States. *Emerging Infectious Diseases* 28(5):940–947. https://doi.org/10.3201/eid2805.212488.

Zdrazílek, J., K. Zácek, J. Chvapil, V. Mikesová, L. Pokorná, V. Tomanová, J. Trauc, and J. Vrábková. 1971. [Virological surveys of the presence of enteroviruses in waste water. I. Incidence of polioviruses in Prague at the end of 1968 and in 1969]. *Ceskoslovenská Epidemiologie, Mikrobiologie, Imunologie* 20(2):67–72.

Zhang, Y., L. Yakob, M. B. Bonsall, and W. Hu. 2019. Predicting seasonal influenza epidemics using cross-hemisphere influenza surveillance data and local internet query data. *Scientific Reports* 9(1):1–7.

Zhang, Y., M. Cen, M. Hu, L. Du, W. Hu, J. J. Kim, and N. Dai. 2021. Prevalence and persistent shedding of fecal SARS-CoV-2 RNA in patients with COVID-19 infection: A systematic review and meta-analysis. *Clinical and Translational Gastroenterology* 12(4):e00343. https://doi.org/10.14309/ctg.0000000000000343.

Zheng, X., Y. Deng, X. Xu, S. Li, Y. Zhang, J. Ding, H. Y. On, J. C. C. Lai, C. In Yau, A. W. H. Chin, L. L. M. Poon, H. M. Tun, and T. Zhang. 2022. Comparison of virus concentration methods and RNA extraction methods for SARS-CoV-2 wastewater surveillance. *Science of the Total Environment* 824:153687. https://doi.org/10.1016/j.scitotenv.2022.153687.

Zuckerman, N. S., I. Bar-Or, D. Sofer, E. Bucris, H. Morad, L. M. Shulman, N. Levi, L. Weiss, I. Aguvaev, Z. Cohen, K. Kestin, R. Vasserman, M. Elul, I. S. Fratty, M. Geva, M. Wax, O. Erster, R. Yishai, L. Hecht-Sagie, S. Alroy-Preis, E. Mendelson, and M. Weil. 2022. Emergence of genetically linked vaccine-originated poliovirus type 2 in the absence of oral polio vaccine, Jerusalem, April to July 2022. *Euro Surveillance* 27(37). https://doi.org/10.2807/1560-7917.ES.2022.27.37.2200694.

# Appendix A

# Committee Members and Staff Biographical Sketches

## COMMITTEE MEMBERS

**Guy H. Palmer (NAM),** *Chair*, holds the Jan and Jack Creighton Endowed Chair at Washington State University (WSU) where he is Regents Professor of Pathology & Infectious Diseases. The founding director of WSU's Paul G. Allen School for Global Health, he leads interdisciplinary health research as the senior director of global health. Dr. Palmer also holds adjunct appointments with the University of Nairobi, the Center for Emerging and Re-emerging Infectious Diseases at the University of Washington, and the Universidad del Valle de Guatemala. He supports the wastewater-based COVID-19 surveillance program in Pullman, Washington, by advising on public health interventions that are informed by wastewater-derived data. Dr. Palmer is a member of the National Academy of Medicine, a Medical Sciences Fellow of the American Association for the Advancement of Science, and a founding member of the Washington State Academy of Sciences, and currently serves on the National Advisory Allergy and Infectious Diseases Council for the National Institutes of Health. He previously served as chair of the National Academies of Sciences, Engineering, and Medicine's Committee on Examining the Long-Term Health and Economic Effects of Antimicrobial Resistance in the United States. Dr. Palmer earned his Ph.D. in infectious diseases from WSU, and his B.S. in biology and a D.V.M. from Kansas State University. He holds honorary doctorates from the University of Bern (Dr. Med. Vet., 2011) and Kansas State University (Ph.D., 2016).

**Ami S. Bhatt** is an associate professor at Stanford University in the Departments of Medicine (Hematology; Blood & Marrow Transplantation) and Genetics. She is a physician scientist with a strong interest in microbial genomics and metagenomics. Her team's research program seeks to illuminate the interplay between the microbial environment and host/clinical factors in human diseases by developing and applying novel molecular and computational tools to study strain-level dynamics of the microbiome, to understand how microbial genomes change over time, and to predict the functional output of microbiomes. She is applying these tools to support wastewater-based disease surveillance in the San Francisco Bay area both for COVID-19 as well as emerging disease. Dr. Bhatt has received multiple awards including the Chen Award of Excellence from the Human Genome Organisation and the Sloan Foundation Fellowship; she is also an elected member of the American Society of Clinical Investigation. Dr. Bhatt cofounded the nonprofit Global Oncology and serves as the director for global oncology for Stanford's Center for Innovation in Global Health. She received her M.D. and Ph.D. from the University of California, San Francisco, followed by residency, fellowship, chief residency, and a postdoctoral fellowship at Harvard Medical School.

**Marisa C. Eisenberg** is associate professor of epidemiology, complex systems, and mathematics at the University of Michigan, and the current director of the Center for the Study of Complex Systems. Her research is in mathematical epidemiology and infectious disease modeling, and blends mathematics, statistics, and epidemiology to understand transmission dynamics, inform intervention strategies, and improve forecasting. Her research includes a range of modeling and analysis of wastewater surveillance for infectious diseases, including polio, SARS-CoV-2, and other pathogens. During the pandemic, Dr. Eisenberg has been closely involved in COVID-19 response efforts at both the state and university levels, developing modeling, analysis, and tools to understand COVID-19 in Michigan, for which she was recently awarded the University of Michigan's President's Award for National and State Leadership. Dr. Eisenberg received her M.S. and Ph.D. in biomedical engineering from the University of California, Los Angeles, and was a postdoctoral fellow at the Mathematical Biosciences Institute at Ohio State University.

Dr. Eisenberg receives U.S. Centers for Disease Control and Prevention Epidemiology and Laboratory Capacity for Prevention and Control of Emerging Infectious Disease funding via the Michigan Department of Health and Human Services to support the state's COVID-19 wastewater surveillance program. Dr. Eisenberg has made public statements regarding the use of wastewater surveillance as a key tool to assess disease beyond COVID-19, including polio.

**Raul A. Gonzalez** is an environmental scientist at Hampton Roads Sanitation District (HRSD). He runs HRSD's molecular pathogen program, which is comprised of a molecular laboratory and field scientists. His group applies molecular methods to manmade infrastructure and adjacent waters. During the COVID-19 pandemic, Dr. Gonzalez led the expansion of HRSD's research and laboratory capacity to identify the best methodological approaches to wastewater-based disease surveillance at the utility. Current projects use nucleic acid-based markers for a variety of applications, including identifying compromised sewer infrastructure and quantifying pathogen removal of various treatment trains. He previously worked at the Los Angeles County Sanitation District. Dr. Gonzalez earned his B.S. in biology from the University of California, Los Angeles, and his Ph.D. in environmental science and engineering from the University of North Carolina at Chapel Hill.

**Charles N. Haas** (NAE) is the L. D. Betz Professor of Environmental Engineering at Drexel University where he has been since 1991. From 2005 to 2020, he was also head of the Department of Civil, Architectural & Environmental Engineering. He also has courtesy appointments in the Department of Emergency Medicine of the Drexel University College of Medicine and in the School of Public Health. For more than 35 years, Professor Haas has specialized in the assessment of risk from and control of human exposure to pathogenic microorganisms and, in particular, the treatment of water and wastewater to minimize microbial risk to human health. He co-directed the U.S. Environmental Protection Agency/U.S. Department of Homeland Security University Cooperative Center of Excellence–Center for Advancing Microbial Risk Assessment. He is a member of the National Academy of Engineering, a Board Certified Environmental Engineering Member by eminence of the American Academy of Environmental Engineers, and a fellow of the American Academy for the Advancement of Science and the American Academy of Microbiology. He has previously served on numerous National Academies of Sciences, Engineering, and Medicine committees, including the Committee on Strategies for Identifying and Addressing Vulnerabilities Posed by Synthetic Biology and planning committees for two symposia on gain of function research with H5N1/H7N9 avian influenza. He received his B.S. in biology and his M.S. in environmental engineering from the Illinois Institute of Technology, and his Ph.D. in environmental engineering from the University of Illinois at Urbana-Champaign.

**Loren P. Hopkins** is the chief environmental science officer at the City of Houston, chief of the Bureau of Community and Children's Environmental Health at the Houston Health Department, and a professor of practice in the Department of Statistics at Rice University. In this capacity, she conducts applied environmental health research and uses the results to inform policies at the City of Houston to improve the health of the community. In May 2020, Dr. Hopkins established a critical wastewater surveillance program at the Houston Health Department that continues to serve as a barometer on the spatiotemporal prevalence of the SARS-CoV-2 virus in the City of Houston. The SARS-CoV-2 wastewater surveillance sampling results have informed public health intervention decisions and aided in improving the response to the COVID-19 pandemic in Houston. She received the 2016 Teaching and Mentoring Award from the Graduate Student Association at Rice University. She holds a B.S. in geological sciences with emphasis in geophysics from the University of Texas at Austin and an M.S. and Ph.D. in environmental science and engineering from Rice University.

Dr. Hopkins is also serving as a paid consultant via StatAnalytics IDD LLC through June 2022 to support work on a U.S. Centers for Disease Control and Prevention Foundation–funded project on best practices in wastewater surveillance. In her capacity as the chief environmental science officer for the City of Houston, Dr. Hopkins has made public statements explaining how the city uses wastewater-based disease surveillance to inform its health initiatives.

**Na'Taki Osborne Jelks** is assistant professor in the Environmental and Health Sciences Program at Spelman College in Atlanta, Georgia, and co-founder of the West Atlanta Watershed Alliance, a community-based environmental justice organization. In her research, Dr. Jelks champions community science and other participatory research approaches; she trains community residents to be watershed researchers who monitor water quality and investigate local environmental conditions, giving them actionable data to press for solutions to urban watershed and community health challenges. In 2021, Dr. Jelks was named an Ecological Society of America Excellence in Ecology Scholar. Since 2018, Dr. Jelks has served on the National Environmental Justice Advisory Council, a federal advisory committee that works to integrate environmental justice into the U.S. Environmental Protection Agency's programs, policies, and activities as well as to improve the environment or public health in communities disproportionately burdened by environmental harms and risks. Dr. Jelks holds a B.S. in chemistry and civil engineering from Spelman College and the Georgia Institute of Technology, an M.P.H. in environmental and occupational health from Emory University, and a Ph.D. in public health from Georgia State University.

**Christine K. Johnson (NAM)** is professor of epidemiology and ecosystem health and director of the EpiCenter for Disease Dynamics at the One Health Institute, University of California, Davis. Her work is committed to transdisciplinary research to characterize impacts of environmental change on animal and human health, inform preparedness for emerging threats, and guide public policy at the intersection of emerging disease and environmental health. Professor Johnson's research pioneers new approaches to characterization of emerging threats and disease dynamics at the animal-human interface in rapidly changing landscapes that constitute "fault lines" for disease emergence, disease spillover, and subsequent spread. She leads the EpiCenter for Emerging Infectious Disease Intelligence, one of the National Institute of Allergy and Infectious Diseases' Centers for Emerging Infectious Disease, to investigate the environment and climate-related drivers for spillover and spread of emerging ebolaviruses, coronaviruses, and arboviruses. She is a member of the National Academy of Medicine and was awarded the Distinguished Scholarly Public Service Award. She earned a B.S. in zoology and political science from Duke University; a Ph.D. in epidemiology from the University of California, Davis; and a V.M.D. in veterinary medicine from the University of Pennsylvania.

**Rob Knight** is the founding director of the Center for Microbiome Innovation and professor of pediatrics, bioengineering, and computer science and engineering at the University of California, San Diego. His research has linked microbes to a range of health conditions, enhanced our understanding of microbes in many environments, and made high-throughput sequencing accessible to thousands of researchers around the world. His laboratory has produced many of the software tools and laboratory techniques that enabled high-throughput microbiome science, including QIIME and UniFrac. He is co-founder of the Earth Microbiome Project; the American Gut Project; and the company Biota, Inc., which uses deoxyribonucleic acid from microbes in the subsurface to guide oilfield decisions. He set up and runs the wastewater COVID-19 detection program and co-founded the COVID-19 testing laboratory at the University of California, San Diego, which performs thousands of clinical tests per day and also sequences viral genomes out of wastewater and clinical samples. He is a fellow of the American Association for the Advancement of Science and the American Academy of Microbiology and received the 2019 National Institutes of Health Director's Pioneer Award and 2017 Massry Prize. Dr. Knight earned his B.S. in biochemistry from the University of Otago and his Ph.D. in evolutionary biology from Princeton University.

**Sandra L. McLellan** is professor in the School of Freshwater Sciences at the University of Wisconsin-Milwaukee. Dr. McLellan's research program studies the human health relevance of microorganisms that flux between the primary habitat of human hosts and environmental reservoirs. Her research has identified new indicators of waterborne disease by characterizing the microbial population structure of sewage and also uses genomics and metagenomics to explore the unique microbiome of urban water infrastructure to understand the complex interactions within the community. Dr. McLellan currently works with the Wisconsin State Laboratory of Hygiene to conduct a statewide SARS-CoV-2 wastewater monitoring program. Dr. McLellan is a fellow in the American Academy of Microbiology and is a member of the International Water Association and the International Society for Microbiology. She received her B.S. in health sciences from the University of Wisconsin-Milwaukee and Ph.D. from the University of Cincinnati College of Medicine in environmental health.

Dr. McLellan receives U.S. Centers for Disease Control and Prevention Epidemiology and Laboratory Capacity for Prevention and Control of Emerging Infectious Disease funding via the Wisconsin Department of Health Services to support the state's COVID-19 wastewater surveillance program.

**Michelle M. Mello (NAM)** is professor of law at Stanford Law School and professor of health policy at Stanford University School of Medicine. She conducts empirical research into issues at the intersection of law, ethics, and health policy, with a focus on understanding the effects of law and regulation on healthcare delivery and population health outcomes. Her research interests include medical liability, public health law, pharmaceuticals and vaccines, biomedical research ethics and governance, and health information privacy. Dr. Mello received the Alice S. Hersh New Investigator Award from AcademyHealth, a Greenwall Faculty Scholars Award in Bioethics, and a Robert Wood Johnson Foundation Investigator Award in Health Policy Research and is a member of the National Academy of Medicine. She holds B.A.s in political science and applied ethics from Stanford University; an M.Phil. in comparative social research from Oxford University, where she was a Marshall Scholar; a J.D. from the Yale Law School; and a Ph.D. in health policy and administration from the University of North Carolina at Chapel Hill.

**John Scott Meschke** is professor and associate chair in the Department of Environmental and Occupational Health Sciences in the School of Public Health at the University of Washington. He is also an adjunct professor in civil and environmental engineering at the University of Washington. Dr. Meschke is an environmental and occupational health microbiologist and virologist specializing in the fate, transport, detection, and control of pathogens in environmental media. Over the past 10 years, his research has focused on development, validation, and implementation of wastewater surveillance methods for poliovirus, salmonella typhi, SARS-CoV-2, and other pathogens. Dr. Meschke is currently serving as a paid consultant to PATH (an international nongovernmental organization) on wastewater surveillance in developing countries and in an unpaid advisory role to the Washington State Department of Health to advise on wastewater surveillance methods in state correctional facilities. Dr. Meschke completed his B.S. in biology and J.D. at the University of Kansas, his M.S. in environmental science at Indiana University, and his Ph.D. in environmental sciences and engineering at the University of North Carolina at Chapel Hill.

Dr. Meschke receives U.S. Centers for Disease Control and Prevention Epidemiology and Laboratory Capacity for Prevention and Control of Emerging Infectious Disease funding, as a subject matter expert, via the Washington State Department of Health to support the state's COVID-19 wastewater surveillance program. Dr. Meschke received donated supplies from Qiagen, Ceres Nano, and Macherey Nagel to evaluate wastewater surveillance products for SARS-CoV-2 without restrictions on the conduct of the research or the publication of the results.

**Rekha Singh** is wastewater surveillance program manager for the Virginia Department of Health where she partners with wastewater treatment plants, laboratories, and local health departments to design and implement the wastewater surveillance program for Virginia. Prior to this, she co-led the State Testing Task Force for pandemic response and led the establishment of the OneLab program to enhance testing in the Commonwealth of Virginia. She has a background in epidemiology, environmental health, contaminants remediation, disease surveillance, water and wastewater, community engagement, the development of novel point-of-use drinking water treatment technologies, and the deployment of these technologies in the real world. Dr. Singh earned a Ph.D. in environmental science and engineering with a focus on environmental health and an M.P.H. in health policy law and ethics from the University of Virginia.

**Neeraj Sood** is professor and vice dean for research at the University of Southern California (USC) Price School of Public Policy and also holds appointments in the Marshall Business School and the Keck School of Medicine at USC. He is senior fellow and director of the COVID-19 Initiative at the USC Schaeffer Center, research associate at the National Bureau of Economic Research, and a visiting scholar at Amazon. He has published several papers on policy responses to the COVID-19 pandemic, including on the effects of shelter-in-place policies on excess mortality, the effects of the pandemic on the use of high-value healthcare, the impact of the pandemic on the use of telehealth, the impact of vaccines on mental health, and the impact of school closures on COVID-19 transmission. Dr. Sood received his B.A. in economics from Delhi University, an M.A. in economics from Delhi School of Economics, an M.A. in economics from Indiana University–Purdue University, and a Ph.D. in policy analysis from RAND Graduate School.

Dr. Sood receives U.S. Centers for Disease Control and Prevention Epidemiology and Laboratory Capacity for Prevention and Control of Emerging Infectious Disease funding via the Los Angeles County Department of Public Health and City of Los Angeles to support the county's COVID-19 wastewater surveillance program.

**Krista Wigginton** is associate professor of environmental engineering in the Department of Civil and Environmental Engineering at the University of Michigan. Her research team focuses on viruses in the environment, including their mechanistic fate, the role they play in urban water microbial ecology, and the development of novel detection methods. Prior to COVID-19, her team published studies on how coronaviruses behave in municipal wastewater and through water treatment processes and developed methods for recovering them from water samples. During the COVID-19 pandemic, she has collaborated on a number of projects focused on implementing wastewater-based epidemiology programs. She is the recipient of a National Science Foundation CAREER Award and the Water Research Foundation Paul L. Busch Award. Dr. Wigginton received her B.S. in chemistry from the University of Idaho, and M.S. and Ph.D. in environmental engineering from Virginia Tech.

Dr. Wigginton is currently working on a U.S. Centers for Disease Control and Prevention (CDC) Foundation–funded project with Verily Life Sciences to demonstrate how to conduct and report wastewater-based disease surveillance. Dr. Wigginton receives CDC Epidemiology and Laboratory Capacity for Prevention and Control of Emerging Infectious Disease funding via the Michigan Department of Health and Human Services to support the state's COVID-19 wastewater surveillance program.

## STUDY STAFF

**Stephanie E. Johnson,** *Study director*, is a senior program officer with the Water Science and Technology Board. Since joining the National Research Council in 2002, she has worked on a wide range of water-related studies, on topics such as desalination, wastewater reuse, contaminant source remediation, coal and metal mining, coastal risk reduction, and ecosystem restoration. Dr. Johnson received her B.A. from Vanderbilt University in chemistry and geology and her M.S. and Ph.D. in environmental sciences from the University of Virginia.

**Alexis Wojtowicz** is an associate program officer for the Board on Population Health and Public Health Practice at the National Academies of Sciences, Engineering, and Medicine. Ms. Wojtowicz has supported the Roundtable on Health Literacy, and consensus studies addressing the health effects of e-cigarettes, research needs in clinical prevention, and clinical guidance regarding per- and polyfluoroalkyl substances exposure. Ms. Wojtowicz has a B.A. in art history from the University of Maryland and an M.P.H. from Johns Hopkins Bloomberg School of Public Health, where she was a fellow of the Bloomberg American Health Initiative with a focus on adolescent health.

**Padraigh Hardin** is a program assistant supporting the Water Science and Technology Board and Board on Earth Sciences and Resources in the Division on Earth and Life Studies. They joined the National Academies of Sciences, Engineering, and Medicine in May 2022. During their last year of undergraduate study at George Mason University (GMU), they conducted research on cloud type and forecast model simulations, which was showcased at the GMU College of Science Research Colloquium. They earned their B.S. in atmospheric sciences from GMU.

# Appendix B

# Board Rosters

## WATER SCIENCE AND TECHNOLOGY BOARD

**CATHERINE L. KLING** (NAS), *Chair*, Cornell University, NY
**NEWSHA AJAMI**, Stanford University, Palo Alto, CA
**PEDRO J. ALVAREZ** (NAE), Rice University, Houston, TX
**JONATHAN D. ARTHUR**, American Geosciences Institute, Washington, DC
**RUTH L. BERKELMAN** (NAM), Emory University, Atlanta, GA
**JORDAN R. FISCHBACH**, RAND Corporation, Santa Monica, CA
**ELLEN GILINSKY**, Ellen Gilinsky, LLC, Richmond, VA
**ROBERT M. HIRSCH**, U.S. Geological Survey, Reston, VA
**VENKATARAMAN LAKSHMI**, University of Virginia, Charlottesville
**MARK W. LeCHEVALLIER**, Dr. Water Consulting, LLC, Morrison, CO
**CAMILLE PANNU**, University of California, Irvine
**DAVID L. SEDLAK** (NAE), University of California, Berkeley
**JENNIFER TANK**, University of Notre Dame, IN
**DAVID L. WEGNER**, Jacobs Engineering, Tucson, AZ

*Staff*

**DEBORAH GLICKSON**, Director
**LAURA J. EHLERS**, Senior Program Officer
**STEPHANIE E. JOHNSON**, Senior Program Officer
**M. JEANNE AQUILINO**, Financial Business Partner

**MARGO REGIER,** Associate Program Officer
**JONATHAN TUCKER,** Associate Program Officer
**EMILY BERMUDEZ,** Program Assistant
**PADRAIGH HARDIN,** Program Assistant
**MILES LANSING,** Program Assistant
**OSHANE ORR,** Program Assistant

## BOARD ON POPULATION HEALTH AND PUBLIC HEALTH PRACTICE

**BRUCE (NED) N. CALONGE** (NAM), *Chair*, University of Colorado, Aurora
**MARCELLA ALSAN,** Harvard University, Cambridge, MA
**ANA V. DIEZ ROUX** (NAM), Drexel University, Philadelphia, PA
**LAURA HERRERA SCOTT,** Summit Health, Towson, MD
**DORA HUGHES,** Centers for Medicare & Medicaid Services, Baltimore, MD
**TAMARRA JAMES-TODD,** Harvard University, Boston, MA
**NICOLA KLEIN,** Kaiser Permanente Northern California, Oakland
**MARGARET MOSS,** University of British Columbia, Vancouver
**JEWEL MULLEN,** University of Texas at Austin
**ANAND PAREKH,** Bipartisan Policy Center, Washington, DC
**THERESE S. RICHMOND** (NAM), University of Pennsylvania, Philadelphia
**JOSHUA SALOMON,** Stanford University, Palo Alto, CA
**MELISSA A. SIMON** (NAM), Northwestern University, Chicago, IL
**SYLVIA TRENT-ADAMS** (NAM), University of North Texas, Fort Worth
**SEAN D. YOUNG,** University of California, Irvine

*Staff*

**ROSE MARIE MARTINEZ,** Senior Board Director
**KATHLEEN STRATTON,** Scholar
**ALINA BACIU,** Senior Program Officer
**ELIZABETH BOYLE,** Senior Program Officer
**GILLIAN BUCKLEY,** Senior Program Officer
**AMY GELLER,** Senior Program Officer
**ANNE STYKA,** Senior Program Officer
**MISRAK DABI,** Financial Business Partner
**ALEXANDRA ANDRADA,** Program Officer
**NEHA DIXIT,** Associate Program Officer
**AIMEE MEAD,** Associate Program Officer
**ALEXIS WOJTOWICZ,** Associate Program Officer
**Y. CRYSTI PARK,** Administrative Assistant
**BRIELLE DOJER,** Research Associate
**KELLY McHUGH,** Research Associate
**ALEXANDRA McKAY,** Research Associate
**DARA ROSENBERG,** Research Associate
**AASHAKA SHINDE,** Research Associate

**MAGGIE ANDERSON,** Research Assistant
**EKENE AGU,** Senior Program Assistant
**AYSHIA COLETRANE,** Senior Program Assistant
**NERISSA HART,** Senior Program Assistant
**GRACE READING,** Senior Program Assistant